METAL OXIDE NANOSTRUCTURES
AS GAS SENSING DEVICES

Series in Sensors

Series Editors: Barry Jones and Haiying Huang

Other recent books in the series:

Compound Semiconductor Radiation Detectors
Alan Owens

Nanosensors: Physical, Chemical, and Biological
Vinod Kumar Khanna

Handbook of Magnetic Measurements
S. Tumanski

Structural Sensing, Health Monitoring, and Performance Evaluation
D. Huston

Chromatic Monitoring of Complex Conditions
Edited by G. R. Jones, A. G. Deakin, and J. W. Spencer

Principles of Electrical Measurement
S. Tumanski

Novel Sensors and Sensing
Roger G. Jackson

Hall Effect Devices, Second Edition
R. S. Popovic

Sensors and Their Applications XII
Edited by S. J. Prosser and E. Lewis

Sensors and Their Applications XI
Edited by K. T. V. Grattan and S. H. Khan

Thin Film Resistive Sensors
Edited by P. Ciureanu and S. Middelhoek

Electronic Noses and Olfaction 2000
Edited by J. W. Gardner and K. C. Persaud

Sensors and Their Applications X
Edited by N. M. White and A. T. Augousti

METAL OXIDE NANOSTRUCTURES AS GAS SENSING DEVICES

G. Eranna

Sensors and Nanotechnology Group
CSIR – Central Electronics Engineering Research Institute
Pilani-333031, India

CRC Press

Taylor & Francis Group

Boca Raton London New York

CRC Press is an imprint of the
Taylor & Francis Group, an **informa** business

A TAYLOR & FRANCIS BOOK

CRC Press
Taylor & Francis Group
6000 Broken Sound Parkway NW, Suite 300
Boca Raton, FL 33487-2742

First issued in paperback 2019

© 2012 by Taylor & Francis Group, LLC
CRC Press is an imprint of Taylor & Francis Group, an Informa business

No claim to original U.S. Government works

ISBN-13: 978-1-4398-6340-4 (hbk)
ISBN-13: 978-0-367-38186-8 (pbk)

Library of Congress Cataloging-in-Publication Data

Eranna, G.
    Metal oxide nanostructures as gas sensing devices / G. Eranna.
       p. cm. -- (Series in sensors)
    Includes bibliographical references and index.
    ISBN 978-1-4398-6340-4 (hardback)
    1. Gas detectors. 2. Metallic oxides--Microstructure. 3. Metallic oxides--Synthesis. 4. Nanostructures. I. Title.

TP159.C46E73 2011
621.381--dc23   2011034377

Visit the Taylor & Francis Web site at
http://www.taylorandfrancis.com

and the CRC Press Web site at
http://www.crcpress.com

Syamala

Raviteja

Keshavaditya

For their continuous support

Contents

List of Tables

Preface

The aim of this book is to develop an integrated micro gas sensor that is compatible with modern semiconductor fabrication facilities so that a small, compact, and low-power device can be created that will be useful in analyzing air ambience with handheld systems. Such miniaturized systems need a good sensor capable of identifying different gaseous species. As of today, there are no specific sensing elements for a specific gaseous species, which makes the task of identifying them even more difficult. The study of metal oxides provides enough indication that they are sensitive to a group of species, and development of such devices is not impossible. Nanostructures add their sensitivity to the existing capabilities of metal oxides as gas-sensing elements.

Chapter 1 provides an introduction to the field. Chapter 2 deals with present-day gas-sensing tools and the necessity for small, miniaturized sensors. Chapter 3 deals with materials and devices. Chapter 4 covers techniques used for gas-sensing applications, such as resistance and capacitance variations. Chapter 5 discusses the issues of sensitivity, concentration, and temperature dependency and the response and recovery times that are crucial for sensors. Chapter 6 presents different techniques used for synthesizing different metal oxides, particularly those having nano-dimensional structures. Gas-sensing properties of nanostructured metal oxides—including aluminum, bismuth, cadmium, cerium, cobalt, copper, gallium, indium, iridium, iron, molybdenum, nickel, niobium, tellurium, tin, titanium, tungsten, vanadium, zinc, and zirconium oxides and mixed oxides—are highlighted in Chapter 7. Chapter 8 deals with active devices that are based on nanostructures. The last chapter, Chapter 9, covers future devices and nanostructured gas-sensor arrays.

I am greatly indebted to Dr. Chandra Shekhar, Director of Central Electronics Engineering Research Institute (CEERI), Pilani, for his keen interest in the subject and for his support. He ensured that the complete infrastructure, necessary network facilities, and access to e-journals were made available for this task. I was fortunate to interact with Dr. Sudhir Chandra, Dr. S.C. Gadkari, Dr. V.K. Jain, Dr. G.V. Ramaraju, Sunil Alag, and Sunita Verma on different issues related to the topic of gas sensors. The artwork for this book was done by R.K. Punia and Rajeev Soni. I also received much support from P.V.L. Reddy, B.C. Joshi, A.K. Gupta, A.K. Sharma, J. Bhargava, B.C. Pathak, P. Kothari, A.K. Singh, B. Lal, H. Sonania, G. Saini, A. Rani, T. Dadhich, T. Mudgal, M. Dwivedi, and A. Balraj at various stages of the manuscript preparation.

Sincere thanks are due to Taylor & Francis for giving me the opportunity to take up this project and also for the support from their entire team. I wish to acknowledge the help from Aastha Sharma and Marsha Pronin.

I would also like to acknowledge the permissions granted by different societies for using their copyright material, particularly Académie des Sciences Elsevier SAS, the American Chemical Society, the American Institute of Physics, Annual Reviews Inc., different Elsevier societies, IEEE, IOP Publishing Ltd., Materials Research Society and Cambridge University Press, Springer Science and Business Media, The Electrochemical Society (ECS), Wiley-VCH Verlag GmbH & Co. KGaA, and Copyright Clearance Center for clearing many of the requests.

Last but not least, I would like to acknowledge the continuous and unending support I received from my family. I would like to thank my wife, Syamala Eranna, and my sons, G. Raviteja and G. Keshavaditya, for waiting patiently for the completion of this book and also for sparing their precious time.

G. Eranna, PhD
Sensors and Nanotechnology Group
Council of Scientific and Industrial Research
Central Electronics Engineering Research Institute
Pilani, India

Abbreviations and Acronyms

1-D	One dimensional
2-D	Two dimensional
3-D	Three dimensional
A/F	Air-to-fuel (mixing)
AAO	Anodized aluminum oxide
ac/AC	Alternating current
ACH	Aluminum chlorohydrate
AFM	Atomic force microscopy
APCVD	Atmospheric pressure chemical vapor deposition
AQM	Air quality monitoring
BET	Brunauer–Emmet–Teller (studies)
BTEX	Benzene, toluene, ethylbenzene, and xylene
CCFET	Capacitively controlled field-effect transistor
CHEMFET	Chemically sensitive field-effect transistor
CMOS	Complementary metal-oxide-semiconductor (device)
CMP	Chemical mechanical polishing
C–V	Capacitance–voltage (characteristics)
CVD	Chemical vapor deposition
CVS	Chemical vapor synthesis
dc/DC	Direct current
DMMP	Dimethyl methylphosphonate [$CH_3PO(OCH_3)_2$]
DSSC	Dye-sensitized solar cell
DTA	Differential thermal analysis
EDS	Energy-dispersive spectroscopy
EDX	Energy-dispersive x-ray fluorescence
EU	European Union
FED	Field emission device
FE-SEM	Field emission scanning electron microscope
FET	Field-effect transistor
FTIR	Fourier transform infrared (spectroscopy)
GB	Grain boundaries
GC	Gas chromatograph
HREM	High-resolution electron microscope
HRTEM	High-resolution transmission electron microscopy
HSGFET	Hybrid suspended gate field-effect transistor
HVTE	High-vacuum thermal evaporation
HWCVD	Hot-wall chemical vapor deposition
HWCVD	Hot-wire chemical vapor deposition

ITO	Indium tin oxide
I–V	Current–voltage (properties)
LASER	Light amplification by stimulated emission of radiation
LED	Light emitting diode
LPCVD	Low-pressure CVD
LPG	Liquefied petroleum gas
MBE	Molecular beam epitaxy
MCT	Mercury-cadmium-telluride (detector)
MEMS	Microelectromechanical system(s)
MOCVD	Metal-organic chemical vapor deposition
MOD	Metallo-organic deposition/decomposition
MOS	Metal-oxide-semiconductor (active device)
MOS	Metal-oxide-semiconductor (capacitor)
MOX	Metal oxides
MS	Mass spectrometer
MWNT	Multi-wall nanotube
NCAFM	Non contact atomic force microscopy
NIL	Nanoimprint lithography
NIOSH	National Institute of Occupational Safety and Health
NW	Nanowire
o-	*Ortho-*
OSHA	Occupational Safety and Health Administration
p	Partial pressure
p-	*Para-*
PBLG	Poly-γ-benzyl-L-glutamate
PCA	Principal component analysis
PECVD	Plasma-enhanced CVD
PEL	Permissible exposure limit
PLD	Pulse laser deposition
PMMA	Polymethylmethacrylate
ppb	Parts per billion
ppm	Parts per million
PVD	Physical vapor deposition
PVP	Polyvinyl pyrrolidone
RBS	Rutherford back scattering
RC	Resistance–capacitance (parameter)
RE	Rare earth (oxides)
REL	Recommended exposure limit
RF	Radio frequency
RGTO	Rheotaxial growth and thermal oxidation process
RH	Relative humidity
SAED	Selected-area electron diffraction
SEM	Scanning electron microscope
SG	Sol–gel (technique)
SGFET	Suspended gate field-effect transistor

SKPM	Scanning Kelvin probe microscopy
STEL	Short-term exposure limit
SWNT	Single-wall nanotube
TBA	Tributylamine
TCO	Transparent conductive oxide
TEA	Triethylamine
TEM	Transmission electron microscopy
UV	Ultraviolet (radiation)
VLS	Vapor–liquid–solid (process)
VOC	Volatile organic compound(s)
VS	Vapor–solid (process)
XRD	X-ray diffraction

Symbols

E_a	Activation energy
θ	Adsorption coverage
k	Boltzmann constant
β	Coefficient of Volkenstein isotherm
m	Constant (relating to defect density)
γ	Constant (relating to material property)
D_c	Crystallite size
L	Depletion layer thickness
w	Depletion region depth
ε	Dielectric constant
σ	Electrical conductivity
Φ	Electrical potential
q	Elemental charge
E_g	Energy bandgap
V_{FB}	Flat-band voltage
N_D	Free carrier density
X	Grain boundary neck diameter
V_B	Grain boundary potential
R_{gb}	Grain boundary resistance
D	Grain size
N_i	Ion density in charge region
λ	Lambda point (automobile ignition)
D_c	Nanocrystal size
f	Occupation probability
p_{O_2}	Oxygen partial pressure
p	Partial pressure (general)

ε_o	Permittivity of vacuum
C	Proportionality constant
β	Rational fraction value (usually 1 or ½)
R	Sensing element electrical resistance
G	Sensing element electrical conductivity
S	Sensor response/sensitivity
V_s	Surface barrier height
N_s	Surface charge
T	Temperature
N_t^-	Trapped charge density

1

Introduction

A chemical gas sensor can be defined as a device, which upon exposure to a gaseous species or molecules, alters one or more of its physical properties, such as mass, electrical conductivity, or dielectric properties, in a way that is possible to measure and quantify. These changes deliver an electrical signal, with a magnitude that is proportional to the concentration of the gas under test, which is measured as a quantity of the gas to which the measuring sensor is exposed. These changes, in some cases, may be spontaneous or they could be a slow response taking several minutes or more. The device should also show a reverse property after the gas has been removed. This reversal may take several minutes to hours depending on the nature of material and the gas involved. However, in certain cases, this reversal may not take place. The former case is ideal for gas-sensing applications. At present, many techniques are used for the purpose of gas detection and each technique has certain advantages and disadvantages as well. Depending on the nature of material, each sensor is known to be sensitive to a group of a family of gases and similarly each gas can be detected by different materials.

Gas sensors have a great influence in many important areas namely environmental monitoring, domestic safety, public security, automotive applications, air-conditioning in aeroplanes, spacecrafts, etc. Air pollution has been one of Europe's main political concerns since the late 1970s. In 1996, the European Council adopted a Framework Directive on ambient air-quality assessment and management (Directive 96/62/EC of September 27, 1996) based on common methods and criteria in the European Union (EU) Member States. A series of "directives" have been issued to control the levels of identified gas pollutants and also to monitor their concentrations in the ambient air under observation [1]. Table A.1 briefs the environmental safety standards for some selective volatile organic compound (VOC) vapors and pollutant gases. National Institute of Occupational Safety and Health (NIOSH) and Occupational Safety and Health Administration (OSHA) directives are summarized for benzene (C_6H_6), ethanol (C_2H_5OH), methane (CH_4), methanol (CH_3OH), and propanol (C_3H_7OH) vapors. Similarly, pollutant gases, such as ammonia (NH_3), carbon dioxide (CO_2), carbon monoxide (CO), hydrogen sulfide (H_2S), nitrogen compounds (NO, NO_2, NO_x), and sulfur dioxide (SO_2), are discussed for their limits of exposure. Recommended exposure limit (REL), short-term exposure limit (STEL), and permissible exposure limit (PEL) parameters are outlined in this chapter. The current EU Policy on Air Quality aims at developing and implementing appropriate instruments to monitor and improve the air quality. European legislation also defines precise obligations, related to the information of the public, in the event of significant pollution epidoses. Besides, any citizen has a right to demand national and local authorities to take action to improve the quality of air. As a consequence, every member country of the EU has been instructed to establish a network of air-quality monitoring (AQM) stations in main cities and to inform its citizens about the air quality on a daily basis.

In recent years, a great deal of research effort has been directed toward the development of small dimensional (miniaturized) gas-sensing devices for practical and field applications ranging from toxic gas detection to pollution monitoring in the ambient living environment [2]. With the increasing demand for better gas sensors of higher sensitivity and greater selectivity, intense efforts are being made in the scientific community to find more suitable materials with the required surface and bulk properties for use in these devices as a sensing element. Detection and quantification of gaseous species in air as contaminants (or pollutants) at a reasonably low cost is becoming increasingly important.

Environmental concerns and health and safety regulations have necessitated an increased use of sensors. Wherever sensor technology has been applied, it has proved to be very beneficial in improving energy efficiency and service, product quality, emission reduction, and in general improved the quality of life. There are many polluting species that have drawn attention. Among these gaseous species, the important species are oxides of nitrogen (NO, NO_2, and NO_x); oxides of carbon (CO, CO_2); hydrogen sulfide (H_2S); sulfur dioxide (SO_2); ozone (O_3); ammonia (NH_3); and many VOCs such as methane (CH_4), propane (C_3H_8), liquid petroleum gas (LPG), acetone (CH_3COCH_3), benzene (C_6H_6), and toluene ($C_6H_5CH_3$). In automotive industry, the demand for more fuel efficiency with lesser emission levels has mandated the need for highly efficient sensors in their exhaust systems [3].

H_2S is one of the toxic and inflammable gases used in large amounts by various chemical industries and research laboratories. It is used as a process gas, in the production of heavy water plants by dual-temperature exchange reaction and also in chemical vapor deposition (CVD) reactors. H_2S gas is formed in the soil by bacterial reduction of sulfates and occurs naturally on earth and flooded grounds. It also exists in hydrocarbon reserves like natural gas and oil. The threshold for H_2S detection by human nose is as low as 0.02 ppm; a person intoxicated with H_2S no longer discerns its characteristic smell before the dangerous threshold values of 50–100 ppm is reached. The occupational exposure limit to H_2S is 10 ppm (for 8 h). Therefore, there is a need to detect H_2S gas when present at a few ppm levels and preferably at sub-ppm levels in the air.

Methane (CH_4) and carbon monoxide (CO) leaks are a danger in any coal-mining operation. Once ore is excavated, it is imperative to block the empty passageways lest they fill up with potentially explosive gases. Explosions are likely when sufficient quantities of oxygen are introduced inside for filling.

A high concentration of methane in the mine means it is a threat for the miners as they get exposed to these gases. These gases deoxygenate the blood by forming methemoglobin and paralyze the brain and destroy the cardiovascular system. By bonding with hemoglobin to form methemoglobin instead, methane deprives the human body of the much-needed oxygen. Monitoring these gases is crucial for the safety of mine and miners.

There are increasing interests in sensing or monitoring of chlorine gas (Cl_2) in automobile exhaust with relation to air pollution due to chlorine-containing compounds such as tetrachloromethane (popularly known as carbon tetrachloride) (CCl_4), chlorobenzene (C_6H_5Cl), dioxin,* and so on. Particularly, to prevent dioxin contamination, the sensors should detect very dilute Cl_2 gas of ppb level in the exhaust. The detection of Cl_2 gas has been reported for the conductivity-type sensors using semiconductor oxide, the potentiometric sensors using solid electrolyte, and the optochemical sensors using dye. However, these sensors could detect Cl_2 gas of only sub-ppm level, except for the optochemical sensor using *o*-tolidine† [4]. Thus, the high-sensitivity Cl_2 gas sensor will be increasingly demanded for exhaust monitoring.

The development of a convenient method for detection of alcohol has always been in great demand in biotechnology and chemical industries. There is an increasing demand on CO_2 sensors in the field of agricultural and biotechnological processes as well as air-conditioning systems or monitoring of the exhaust gases. Relative humidity is an indication of the amount of water vapor in air as a percentage of the total amount possible at a given temperature and the presence of this humidity has its influence on the functioning of these gas sensors.

Formaldehyde (HCHO) is used widely by industries to manufacture building materials and numerous other household items. It is also a by-product of combustion and certain other natural processes, as a result of which, it may be present in substantial concentrations both indoors and outdoors. At room temperature, it is colorless with a pungent and irritating odor. It is highly reactive, readily undergoes polymerization, is highly flammable, and forms explosive mixtures in air. HCHO is emitted into indoor air by the adhesive portion that binds pressed wood products. As per U.S. Consumer Safety Commission 0.1 ppm levels can cause burning sensation in the eyes, nose, and throat; nausea; coughing; chest tightness; wheezing; skin rashes; and allergic reactions. Studies have shown that HCHO emission may cause cancer in humans in the long run.

Physics, chemistry, and technology of sensors require a better understanding of both the bulk and surface properties of the gas-sensing materials. One expects that a sensing element should have high sensitivity and selectivity

---

* Dioxin is a heterocyclic organic antiaromatic compound with the chemical formula $C_4H_4O_2$. It exists in two isomeric forms: 1,2-dioxin (*o*-dioxin) and 1,4-dioxin (*p*-dioxin).
† Tolidine is a group of isomeric organic compounds and the most prevalent is *o*-tolidine (also known as 2-tolidine). This is a commercially important aromatic amine mainly used for dye production and for certain elastomers.

for the test gas, small response and recovery times, no environmental degradations during its usage, preferably at room temperature operation, low-power consumption, and without using expensive noble metals/materials for their fabrication. The method of preparation of the sensor material plays an important role in manifesting the sensor's electrical characteristics. In many of the sensors, the sensing material's surface generally oxidizes the sensing gas either fully or partially. These surface reactions and their properties on surface play a key role in identifying the sensing material and the specific gas to which it is sensitive. However, there is no single gas sensor existing today that is selective to a particular gas.

Presently, many analytical techniques are used for sensing the gases. They include Fourier transform infrared (FTIR) spectroscopy, which uses infrared spectral characteristics of gaseous species; gas chromatographs (GC), which use analytical columns to separate the species; and mass spectrometers (MS), which identify molecules through characteristic variable deflection, depending on the mass and charge, from the applied magnetic field. The spectroscopic technique has been successfully used to investigate the nature of surface species adsorbed.

The main emphasis of this book is on gas-detection techniques using metal oxide (MOX) nanomaterials by observing the variation of their physical properties and their measurements, estimating the air quality in the environments we live and the safety issues associated with them. At present the research on these gas sensors is aimed at obtaining new materials to achieve highly sensitive and selective with long-term operating devices. For environmental studies, there is a demand for battery-operated devices and portable units for air-quality assessment.

Gas sensors in the form of thin or thick films seem to be more promising detectors over the pellet form, because they are of low cost, are rugged, and utilize low power for their operation. Even though the semiconducting oxides used for gas sensors are active, for sensing applications, a small amount of a catalytic metal or MOXs are often added to improve the selectivity and sensitivity of the sensing element. In practical applications, till now oxidation catalysts employed for sensing have a transition metal or a noble metal supported on these MOXs. The thin film form is justified for technological reasons because it is possible to prepare devices with small size that can easily be integrated in an array form. Different deposition methods like physical vapor deposition (PVD), CVD, and sol–gel (SG) techniques are used for the preparation of these thin films. Sputtering is a preferred technique over evaporation in many applications due to a wider choice of materials to work with, better step coverage offered, and better adhesion to the substrate platform. The target material to be deposited is sputtered away mainly as neutral atoms by momentum transfer and the ejected surface atoms are deposited onto the substrate placed on the anode in a reactor chamber. Sputter techniques (including DC, RF, and magnetron sputtering) are currently employed in both laboratories and production settings, whereas thermal vacuum evaporation mainly remains one of the laboratory tools at present. However, the thermal evaporation technique is not recommended for MOXs.

The other PVD techniques reported for MOXs include cathodic vacuum arc deposition, condensation from vapor phase, controlled solid–vapor process, direct metal heating in oxygen ambient, focused electron beam evaporation, laser ablation, molecular beam deposition, sublimation of metal, and ultrasonic spray pyrolysis. Similarly, different approaches are possible in the chemical route. They include aerosol technique, anodization in chemical environment, atmospheric pressure CVD (APCVD) technique, chemical gas phase reaction, chemical synthesis, controlled hydrolysis, different aqueous techniques, electrochemical route, hot-wire CVD technique, hydrothermal solution treatment method, low-pressure CVD (LPCVD) technique, plasma-enhanced CVD (PECVD) technique, SG technique, solvothermal treatment process, spray precipitation technique, template-directed hydrothermal process, and wet chemical route.

Due to the huge application range, the need of cheap, small, low-power consuming, and reliable solid-state gas sensors have grown over the years and triggered a worldwide research to overcome the shortcomings presently felt. The science and technique of growing different types of nanomaterials has become the flavor of the day with research being carried out by different academic and research institutions all over the world for better and more versatile materials and nanomaterials for specific applications. Among the nanostructured materials, application of nanocrystals, nanowires, nanobelts, and nanotubes constitute the major categories in the two-dimensional (2-D) and three-dimensional features. Both carbon-based and inorganic-based nanostructures are the major areas where research activities are being carried out at present. In the past several years, different nanostructures of carbon and inorganic materials were studied for various applications that included MOXs, nitrides, carbides, and chalcogenides. Among the inorganic nanostructures the main concentration is on the MOX nanostructures that are highly useful for the purpose of gas-sensing applications.

A nanotechnology-based breathing sensor could tell whether the person has Type-1 diabetes or not by analyzing their breath. This is based on the analysis of diabetic ketoacidosis, a potentially serious complication that happens when diabetics do not take enough insulin. If they have ketoacidosis, there is a dangerous buildup of acetone in the blood and they exhale larger amounts of acetone. By testing the levels of acetone, it is easy to estimate the condition of the person.

At present, major research challenges in this field of gas (or vapor)-sensing specialization are (i) VOC characterization against complex backgrounds environments (e.g., BTEX [benzene, toluene, ethylbenzene, and xylene VOCs], landfill conditions, and indoor and cabin air quality), (ii) identification of normal and abnormal variations in the gaseous markers of disease in breath and gut gases for medical diagnostics, (iii) improved selectivity and stability for semiconductor and nanomaterial gas sensors at ppb concentration levels, (iv) combinatorial methodology for optimizing sensing materials for e-nose development, and (v) integrated microelectromechanical systems (MEMS) using miniaturized sensing arrays with suitable electronics for widespread applicability.

2

Miniaturized Solid-State Gas Sensors

2.1 Gas Chromatography–Mass Spectrometry

The use of a mass spectrometer as the detector in gas chromatography was developed in the 1950s by Gohlke and McLafferty [5,6]. This is a chemical analysis instrument for separating gaseous species in a complex sample. The system uses a flow-through narrow tube, known as the column, through which different gaseous species of a test gas pass in a stream at different rates depending on their various chemical and physical properties. As the gaseous species exit at the end of the column, they are detected and identified electronically. The stationary phase in the column is to separate different species by causing each one to exit the column at different time intervals. Since each type of molecule has a different rate of progression, the various components of the analyte mixture are separated as they progress along the column and reach the end of the column at different times (retention time). A detector is used to monitor the outlet stream from the column; thus, the time at which each component reaches the outlet and the amount of that component can be determined.

Gas chromatography–mass spectrometry (GC–MS) is used to identify different substances within a test sample and is highly useful for drug detection, fire investigation, environmental analysis, explosives investigation, and identification of unknown samples. These are bulky, fragile, and originally limited to laboratory settings. However, the working principle and species identification are authentic in nature and analyzing unknown gaseous species is accurate.

2.2 Fourier Transform Infrared Spectrometer

The foundations of modern Fourier Transform Infrared Spectroscopy (FTIR) were laid in the latter part of the nineteenth century by Michelson and Raleigh who recognized the relationship of an interferogram with its spectrum by a

Fourier transformation [7]. It was not until the advent of computers and the fast Fourier algorithm that interferometry began to be applied to spectroscopic measurements. Transmittance spectroscopy is primarily used to detect certain impurities or specific bonds present in different gases and vapors.

There are two ways to monitor the concentrations of trace gases in the atmosphere. The first is to draw the atmosphere in the region of interest into a long-path gas cell and the second is to measure the spectrum of the atmosphere *in situ*.

FTIR has three advantages for monitoring of toxic air pollutants over point sampling techniques such as GC–MS or extractive FTIR monitoring: versatility, time, and safety. FTIR is capable of measuring any compound that exists at high enough concentration and has an IR signature. Of the 189 hazardous air pollutants listed in the U.S. Clean Air Act and its amendments, 135 pollutants and an even larger number of compounds for which environmental or health concerns have been raised can be monitored. Libraries of the air-broadened reference spectra of over 1000 vapor phase compounds are available [8]. FTIR provides the ability to examine all these compounds without modification of the analysis conditions.

FTIR instruments may be used indoors or outside and can sample areas where it would be impractical to place other sampling devices. Path lengths anywhere from 1 m to more than 1 km in length can be sampled, although the typical distance from retroreflector to spectrometer is between 100 and 200 m. There is a real benefit in measuring the spectra of VOCs with unresolvable rotational fine structure by measuring the spectrum at lower resolution than 2 cm^{-1}. For small molecules with resolvable rotational fine structure, the effect of resolution on analytical accuracy depends on the presence of other analytes with nearby absorption bands and the presence of water lines in the same spectral region.

One of the major experimental difficulties with FTIR measurements is the need to minimize the effect of water vapor and carbon dioxide on the recorded spectra. With path lengths of over 100 m, essentially all radiation between about 4000 and 3500 cm^{-1} and 2000 and 1300 cm^{-1} is totally absorbed by atmospheric water vapor, and the radiation between 2150 and 2500 cm^{-1} and between 700 cm^{-1} and the cutoff of the mercury-cadmium-telluride (MCT) detector is lost because of absorption by CO_2. Thus, useful analytical information can be obtained from FTIR spectra only in the three atmospheric windows (700–1300, 2000–2150, and 2500–3000 cm^{-1}). Even in these atmospheric windows, however, water lines can still be observed in the spectrum.

2.3 Miniaturized Solid-State Gas Sensors

The present day analytical instruments provide fairly accurate information to detect gaseous species. This type of instruments are proven to be accurate to detect explosives at airports, drug abuse screening, and analyzing air pollutants in the environment. However, these instruments require skilled and knowledgeable operators [9]. They are also very expensive to maintain and are more suitable for laboratory tabletops, specific online applications, and in-plant installations monitoring the general atmosphere prevailing therein. They are bulky in size and are used as a last resort for applications where no suitable gas-sensing device is readily available. For field applications, this type of instruments is difficult to handle and the alternative is to develop battery-operated units with miniaturized sensor-detection systems.

The first solid-state semiconductor gas sensor was introduced in 1968 by Taguchi for the detection of hydrocarbons [10]. In 1972, International Sensor Technology introduced a gas sensor for the detection of hydrogen sulfide (H_2S) gas. Subsequently, many more gas sensors were introduced to detect various hazardous gases at ppm-level concentrations. There are hundreds of different gases present in an environment under study, and it is always difficult to identify each one of them. Depending on the application, one may require the detection of a specific gas, while eliminating the rest of the gases as background. Others may require complete information about different gases present in the environment under study.

The development of miniaturized sensors and associated sensor systems is an interdisciplinary task, which requires detailed and expert knowledge from completely different research fields. Generally, computer scientists, electrical and electronic engineers, physicists, chemists, or medical experts may be involved in bringing out a useful gas sensor. It is visualized that metal-oxide-based solid-state gas sensors present a promising picture for the development of integrated sensor. This is because many of the key gaseous species can be detected by suitably modifying the sensing element of these metal oxides for the purpose. Furthermore, many of these materials can be deposited, in thin-film form, and this type of depositions fit well into the existing microfabrication techniques. It is easy to manufacture microelectronic sensor structures with materialistic features of thin layers. This approach presents specific advantages, such as low cost for the mass production, facilities to produce, possibility to manufacture sensor arrays on a single chip, and possibility to guarantee a good reproducible sampling. A handy sensor system is an added advantage for field, urban, and remote area applications unlike presently used bulky systems that are more suited for fixed locations.

In general, the MEMS technology does not match with the CMOS technology due to the restrictions posed by device fabrication and usage of certain chemicals. As an example, serious limitations are implied if the "release etch," of suspended platforms, is performed after aluminum contact pads that are already realized on the process-bound silicon wafer. Several research groups [11] have demonstrated silicon-processing-compatible, simple, and low-cost method for processing of thermally isolated suspended membranes to make sensor fabrication more compatible with standard CMOS processing. Thin suspended platforms in MEMS structures pose other practical problems especially during the final stages of wafer dicing of fully processed MEMS device with exposed device regions.

Many of the sensor systems exhibit hysteresis behavior mainly due to material property used for detection. However, the systems based on infrared and photoionization instruments do not exhibit this property, but many other sensors, including electrochemical, solid state, and catalytic sensors do exhibit hysteresis. This leads to a range of detection and it is an inherent property of these sensors and difficult to avoid.

3

Gas-Sensing Materials and Devices

More than 50 years ago, Brattain and Bardeen [12] discovered that gas adsorption onto a semiconductor produces a change in the electrical conductance of the material. From that time onward, a detailed study on semiconducting materials was carefully done with a variety of gaseous species that brought about electrical changes. A great amount of research was carried out in order to realize commercial semiconducting devices based on this principle. The first proposal for this type of devices came in 1962 by Seiyama et al. [13] and a patent by Taguchi [10]. From this point onward, many more devices came into existence for a variety of applications. Since these gas sensors utilize the chemical sensitivity of material surfaces for sensing applications, surface behavior became a point of focus for many researchers. This type of gas sensors have been investigated worldwide for development of new applications as well as for progress of basic understandings of these sensors. Researchers have contributed to piling up a great amount of practical as well as basic information about sensing materials and sensing mechanisms. However, semiconductor gas sensors still remain far from being understood satisfactorily. At present, some commercial devices are available and they mainly use electrical conductance variations of materials for detecting different gaseous species in the environment.

Chemiresistors are the simplest of the chemical sensor technologies available for miniaturization into field and portable instruments. Chemiresistors possess advantages of compact size, simple fabrication, low cost, and simple measurement electronics. The other sensor technologies are the potentiometric chemically sensitive field effect transistors (CHEMFETs) and complex solid-state sensors [14]. The CHEMFETs are inherently more complex to fabricate and require more extensive control and measurement electronics.

Since these semiconductor gas sensors are based on the principle that their electrical properties change in the presence of different gaseous environments, it has been found that the adsorption of atomic and molecular species on inorganic semiconductor surfaces can affect surface properties not only related to the electrical conductivity but also influence the surface potential. Careful study of these variations in the ambient atmosphere and their application in gas-detection scheme, through the development of chemo-resistive semiconductor gas sensors, opened up a new field of study [15].

Solid-state semiconductor based gas sensors are more versatile sensors, as they detect a wide variety of gases, and can be used in many different applications. Different response characteristics are achieved by varying the semiconductor materials, processing techniques, and suitable sensor-operating temperatures. Though these sensors are susceptible to other interfering gases, the main strength is their long life, approximating 10 years of useful life. Among the unique attributes of these sensors is the ability to detect very low as well as high concentration levels. In one way, this simplifies the system design and also the maintenance of these sensors.

Selectivity has always been a major hurdle for solid-state gas sensors. There have been a number of approaches developed to improve the selectivity of gas sensors, including doping metal impurities, using impedance measurement, modulating operating temperature, surface coating, etc.

The development of gas-sensor devices with optimized selectivity and sensitivity has been gaining prominence in recent years. Because of their simplicity and low cost, semiconductor metal-oxide gas sensors stand out among the other types of sensors. Sensitivity in these types of sensors is generally enhanced either by appropriate doping of an element, which modifies the carrier concentration values or charge mobility, or by microstructural changes, such as reduction of the oxide particle grain size [3]. Nanostructured materials present new opportunities for enhancing the properties and performance of gas sensors because of the much higher surface-to-bulk ratio compared to coarse-grained materials.

4

Metal-Oxide-Based Gas-Sensor Devices

Sensing behavior is the most important and well-known property of metal-oxide materials. In addition to the sensitivity to light (photon) energy and external pressure, metal oxides demonstrate high sensitivity to their chemical environment. With the capacity to operate in harsh environment, they surpass other chemical sensors in their sensitivity, reliability, and durability. A sensor detects a change in the gas atmosphere due to a change in the electrical properties of sensing elements. It was found that a catalytic reaction and a gas-sensing process on a metal oxide are analogous to each other as both processes involve surface adsorption and chemical reaction with the surrounding gas environment.

Metal oxides are being investigated for gas-detection applications. This is because of their working mechanism being the variation of surface electrical conductance in the presence of a gaseous environment. This induced electrical conductivity is altered by the adsorption of gaseous species from the ambient atmosphere. The metal oxide causes the gas to dissociate into charged ions or complexes on the surface, which results in the transfer of electrons. This property has been exploited and used for the detection of inflammable and toxic gases. The electrical characteristics of these materials critically depend on the chemical composition of the metal-oxide surface and also on the nature and also on the work function behavior. The nature of the surface sites, the electron donor/acceptor properties of the test gas, the adsorption coefficient, the surface reactions, and subsequent desorption of the gases are the key features that determine the performance of semiconductor gas sensors. The interaction between the gas and solid mainly takes place on the surface, and hence the amount of atoms residing in grain boundaries (GBs) and the interface is critical for controlling the properties of these sensors [16]. Generally, these gases induce a resistance increase on n-type semiconductor oxides. This resistance increase can be associated with surface processes and explained in terms of capturing free electrons from the conduction band of the semiconductor by the adsorbed species [17]. It is generally recognized that in metal-oxide samples, the air-conductance behavior mainly represents the influence of surface reactions including electron exchange with the oxide, most notably those involving atmospheric oxygen and water vapor.

The sensing mechanism in metal-oxide gas sensors is related to ionosorption of species over their surfaces. The most important ionosorbed species when operating in ambient air are oxygen and water. The sensing mechanism is mainly governed by the fact that the oxygen vacancies on the oxide surfaces are electrically and chemically active. In this case, two kinds of sensing responses have been observed. (1) Upon adsorption of charge-accepting molecules, such as NO_2 and O_2, at the vacancy sites, electrons are withdrawn and effectively depleted from the conduction band, leading to a reduction of conductivity. (2) On the other hand, in an oxygen-rich environment, gas molecules such as CO and H_2, could react with the surface-adsorbed oxygen and consequently release the captured electrons back to the channel, resulting in an increase in conductance [18]. Conclusively, if one categorizes such redox-sensing response into reducing and oxidizing, which manifests in an increase and a decrease in the channel conductance, the sensing responses can be represented using two examples:

$$\text{Reducing response:} \quad CO + O \rightarrow CO_2 + e^-$$

$$\text{Oxidizing response:} \quad NO + e^- \rightarrow NO^-$$

To avoid long-term changes metal-oxide-based gas sensors, should be operated at temperatures low enough so that appreciable bulk material variation never occurs and is high enough so that gas reactions occur in a time on the order of the desired response time. When a metal oxide is operated in the high-temperature ranges, the charge-transfer process, induced by surface reactions, determines its electrical resistance changes.

The fundamental sensing mechanism of these sensors lies, due to the interaction process, between the surface complexes such as O^-, O_2^-, H^+, and OH^- reactive chemical species and the gas molecules to be detected. The effects of the microstructure, namely, ratio of surface area to volume, grain size, and pore size of the oxide particles, as well as film thickness are well recognized. Lack of long-term stability has until today prevented wide-scale application of this type of sensor. The most recent research has been devoted toward nanostructured oxides, since reactions at GBs and complete depletion of carriers in the grains can strongly modify the material transport properties. Unfortunately, the high temperature required for the surface reactions to take place induces a grain growth by coalescence and prevents the achievement of stable materials. This high temperature ultimately decides the operation of these sensors.

Conductometric oxide thin films are the most promising devices among solid-state chemical sensors, due to their small dimensions, low cost, low-power consumption, online operation, and high compatibility with microelectronic processing. The present application field spans from environmental monitoring, automotive applications, air-conditioning in aeroplanes, spacecrafts, and houses, to sensor networks. The progress made on silicon technology for micromachining and microfabrication foreshadows the development of low-cost, small-sized, and low-power consumption devices, suitable to be introduced in portable instruments and possibly in biomedical systems [19]. More versatile sensors based on nanometal oxides are expected in the near future.

4.1 Techniques Used for Gas Sensing

The surface of a semiconductor bounds the periodicity of the crystal lattice points leading to localized surface states that could be shown by a quantum mechanical analysis. In case the energy levels are lying in the forbidden energy region of the semiconductor material, these surface states can either capture or inject charge carriers into the bulk crystal [20]. The corresponding surface charge induces a counter charge in the surface region built by ionized donors or acceptors present in the crystal.

The charge double layer, which is composed of the negative surface charge and the positive counter charge in the bulk, is equivalent to a band bending of the energy levels in the crystal as can be shown by solving the Poisson equation:

$$\frac{\partial^2 \Phi}{\partial x^2} = \frac{q}{\varepsilon \varepsilon_0} N_i \qquad (4.1)$$

with
Φ as the electrical potential
N_i as the ion density in the charge region at the semiconductor surface region

The Schottky approximation leads to

$$V_s = \frac{q N_s^2}{2 \varepsilon \varepsilon_0 N_i} \qquad (4.2)$$

which describes the relation of the potential difference between the surface and the bulk as a function of the surface charge N_s.

If the semiconductor is exposed to any gaseous atmosphere, additional gas-induced surface states arise due to the adsorption of ambient gas species very close to the surface region. This is shown in Figure 4.1a that shows the mechanism for the adsorption of oxygen species, as an example, which is a common component in most gas-sensing applications. Oxygen adsorbs on the surface by capturing an electron from the conductance band forming an O_2^- ion. This electron capturing is equivalent to an occupation of a localized adsorbance (O_2)-induced surface state. This mechanism is possible because

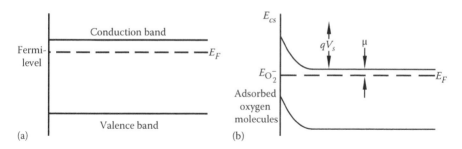

FIGURE 4.1
(a) Simple energy band diagram showing conduction band, valence band, and Fermilevel. (b) Band bending due to adsorption of oxygen species on the sensor surface altering its properties.

the energy level of the surface state is below the Fermilevel of the semiconductor without the presence of oxygen. The charge transfer induces a surface charge and a corresponding band bending in the semiconductor. The Fermilevel is lowered until so much oxygen has adsorbed so that the surface energy level is equal to the semiconductor Fermilevel [20]. This process is called "Fermilevel pinning." Due to the band bending, the surface region is exhausted of free charge carriers thus forming a region of high ohmic resistance, due to charge depletion, for the electrons.

Reducing gases, like CO, get oxidized, to CO_2, consuming oxygen from the sensing material surface region. A dynamic equilibrium between the rate of desorbing oxygen due to the oxidation of reducing gas components and the rate of adsorbing oxygen of the ambient air is formed and the oxygen concentration on the sensor surface gets reduced. Following Equation 4.2, the energy bending is lowered by a certain amount depending on the concentration of the reducing gases. In most cases, the sensor material has a polycrystalline structure, as is shown in Figure 4.1b. All the polycrystalline grains are attached to each other via the GBs. Individual grains are surrounded with neighboring grains. Contact region of grain, with their high ohmic regions, forms a so-called double Schottky energy barrier for the charge carriers, that is, electrons. If we assume thermal emission as the dominant charge-transport mechanism, in this case, the conductivity G over such barriers depends on the energy barrier and is generally given as $G \sim e^{-qV_s/kT}$.

The electrical response signals of any polycrystalline metal-oxide-based sensors may be formally described by resistance (with $R = G^{-1}$) and capacitance units in RC. They contain high concentration of GBs between nanometer-sized crystalline grains. Their sizes and the corresponding size distribution vary with preparation techniques, deposition conditions, annealing temperature, and time durations. Of particular importance for understanding the electrical response signals of nanocrystalline metal-oxide thin films, upon exposure to reducing or oxidizing gases, are measurements of impedances in comparison with dc conductances. The latter are performed with two- or four-point contact arrangements. This makes it possible to separate specific contributions

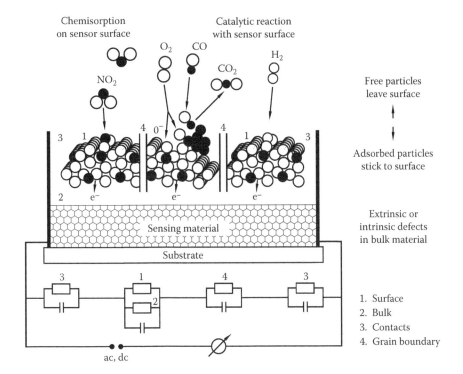

FIGURE 4.2

Schematic representation of elementary steps during detection of molecules. Surface and bulk reactions lead to changes of overall dc or ac conductance of the sensing material, which contain contributions from (1) surface, (2) bulk, (3) boundary or contacts, and (4) grain boundaries. Equivalent circuits with different R–C units, each of which corresponds to a characteristic charge-carrier transport and relaxation processes. Usually two ohmic contacts are used in conductance sensors. In Schottky-diode sensors, one contact/oxide interface is a Schottky barrier with nonlinear current–voltage behavior that is sensitively influenced by chemisorbed particles. (Reprinted from *Sens. Actuat.*, B24–B25, Schierbaum, K.-D., Engineering of oxide surfaces and metal/oxide interfaces for chemical sensors: Recent trends, 239–247, Copyright 1995, with permission from Elsevier Science S.A.)

from contacts, bulk, surface, and GBs, all of which contribute to the sensor response [21]. The details are indicated in Figure 4.2. The variation of these parameters will give rise to a "signal" from the sensing element. The figure shows the schematic representation of elementary steps during detection of molecules. Surface and bulk reactions lead to changes of overall dc or ac conductance of the sensing material, which contain contributions from surface, bulk, boundary, or metal contacts, and finally from GBs. Equivalent circuits with different R–C units, each of which corresponds to a characteristic charge-carrier transport and relaxation processes are indicated in it. Usually two ohmic contacts are used in conductance sensors. In Schottky-diode sensors, one contact/oxide interface is a Schottky barrier with nonlinear current–voltage behavior that is sensitively influenced by chemisorbed particles.

Metal-oxide-based sensors and their corresponding sensor response signals upon interaction with gas molecules may be classified according to their electrical characteristics, that is, the voltage dependence of conductances G and/or capacitances C. This is schematically shown in Figure 4.3, contact arrangements are also shown here. Figure 4.3a shows resistor-type sensors with ohmic properties, which show linear $I–V$ curves. In this case, their conductance $G = dI/dV$ is independent of the applied voltage and reversibly change as a function of partial pressures of gaseous species. Here, p_1 and p_2 are taken as two different partial pressures. Thin film, thick film, pellets, and beads are suitable for such studies. The second type of device is a diode-type sensor, which shows partial-pressure-dependent nonlinear $I–V$ and $C–V$ curves. This typically induces change in cut-in voltage of the diode and dynamic resistances. The third type device is a capacitor-type sensor based on metal-oxide-semiconductor and metal-insulator-semiconductor (M-O-S and M-I-S capacitor) devices, which show voltage-dependent $C–V$ curves. Here, changes in the partial pressure lead to a shift of the flat-band

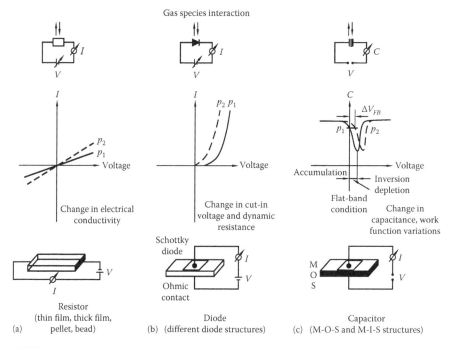

FIGURE 4.3
Current–voltage ($I–V$) and capacitance–voltage ($C–V$) curves of typical metal-oxide-based chemical sensors. (a) Resistor-type sensors with ohmic properties show linear $I–V$ curves, which reversibly change as a function of partial pressure of molecules. (b) Diode-type sensors show partial-pressure-dependent nonlinear $I–V$ curve. (c) Capacitor-type sensors based on MOS devices show voltage-dependent $C–V$ curves. (Reprinted from *Sens. Actuat.*, B24–B25, Schierbaum, K.-D., Engineering of oxide surfaces and metal/oxide interfaces for chemical sensors: Recent trends, 239–247, Copyright 1995, with permission from Elsevier Science S.A.)

voltage ΔV_{FB}, as determined by the transition between charge accumulation and depletion in the semiconductor close to the surface region. The variations of *I–V* and ΔV_{FB} are shown in Figure 4.3b and c [21]. Reliable chemical sensors require the control of geometric and electronic structures of the different surfaces and interfaces at the atomic scale. This concerns in particular the metallic surface dopants of oxide surfaces and their electrical contacts.

In solid-state sensors, the gas molecules adsorb onto the sensing material surface bringing a change in the resistance value of it. In the majority of sensors, the initial output of the sensor is linear or very close to it, but as the gas concentration increases, the output signal gradually reduces and drifts toward the nonlinearity region, resulting poor resolution. After certain concentration, the nonlinearity factor disappears and the signal saturates to a fixed value independent of exposed gaseous concentration. When the gas disappears, the sensor returns to its original condition. In this process, no material consumption takes place and hence provides a long life expectancy in the usage. This is one major advantage of these gas sensors.

The adsorption coverage of the' semiconductor surface is given by the Volkenstein isotherm [22]:

$$\theta = \frac{\beta p}{1 + \beta p} \tag{4.3}$$

where
  θ is the adsorption coverage (ratio between number of chemisorbed species to the total chemisorption sites, N_{ad})
  β is the coefficient of the Volkenstein isotherm
  p is the partial pressure of the gas

The coefficient depends on the position of the Fermilevel at the surface. The existence of a strong form of chemisorption implies the presence of a surface charge given by $N^- = N_{ad}\theta f$, where f is the occupation probability of the strong form of chemisorption given by the Fermi–Dirac statistics.

4.1.1 Resistance Variations

The easiest way to transduce the gas sensor is by simply measuring the dc resistance of the sensing element as a function of the surrounding atmosphere. The design of thin-film resistors normally requires knowledge of the resistance of a certain geometric pattern of the thin-film material between two terminating contacts [23]. The transducer has to provide at least two electrical contacts on the sensing element to measure conductance changes and a heating system in order to maintain the element at the suitable operation temperature. Depending on the material this value is, in general, of the order of a few hundred degrees centigrade. The simplest transducer is a bulk insulating substrate with electrical contacts on one side of the substrate and a heater preferably on the backside of the sensing element.

The relation between the conductance and the oxygen partial pressure is

$$\sigma = C \exp\left(\frac{-E_a}{kT}\right) p_{O_2}^{1/m}$$

(4.4)

where
  σ is the conductivity
  C is proportionality constant
  E_a is the activation energy for the bulk conduction
  p_{O_2} is the oxygen partial pressure [24]

The constant m, the oxygen sensitivity, depends on the defects involved in the conduction mechanism. When the defects are represented by doubly ionized oxygen vacancies, $m = -6$, but if the defects are represented by metal vacancies, different values of m are found.

At constant partial pressure of oxygen and working temperature, the experimental dependence between the conductance and the partial pressure of a single gas is given by

$$G = G_0 + \gamma(p_{gas})^m$$

(4.5)

where
  p_{gas} is the gas partial pressure
  γ is a *constant* that depends on the material

It can be seen that by considering a simplified model of the sensor in which it was formed by parallel chains of grains between the electrodes, the sensor resistance was mainly determined by the grain boundary resistance R_{gb} (resistance between grains):

$$R_{gb} = R_0 \frac{e^{qV_s/kT}}{N_D} = R_0 e^{(E_C - E_F)_{bulk}/kT} e^{qV_s/kT}$$

(4.6)

which is a function of the free carrier density, N_D, and surface barrier height, V_s. So, resistance variations can be produced mainly by the modification of the density of free electrons (movement of the Fermilevel), or by the modification of the barrier height through the change of the charge state of the interfaces of the grains [25]. In the case of solid-state semiconductor gas sensors, the gas-sensitive layer usually consists of a polycrystalline metal-oxide film. The gas-detection principle is based on the variations of the depletion layer at the GBs, in presence of reducing or oxidizing gases, which lead to modulations in the height of the energy barriers for free charge carriers, thus leading to a change of the conductivity of the sensing material. The majority of these devices play a key role on the behavior of the material used, grain size of the polycrystalline, and the dopants added (present) in them.

4.1.2 Capacitance Variations

The capacitance variation property for gas-sensing applications has not gained much popularity, in the form of metal-oxide-semiconductor (MOS) device, to study voltage-dependent $C–V$ characteristics. The shift of the flat-band voltage ΔV_{FB}, as determined by the transition between charge accumulation and depletion in the semiconductor close to the surface region is still not popular in the case of gas-sensing devices. Probably, the exposed regions of the material in MOS capacitor configuration, available surface area of sensing element limit the gas–surface interaction, which is crucial for the sensor.

4.1.3 Electrochemical Techniques

Gas sensing, using mixed potential devices, has been extensively studied over the last two decades. These types of electrochemical techniques are popular for high-temperature gas sensors that develop a mixed potential response offer a simple and inexpensive technique to measure gas concentration in the ambient. Electrokinetic and heterogeneous catalysis measurements are necessary to completely describe the response behavior of a given mixed potential sensor.

The earliest usage of sensors based on the electrochemical techniques goes back to the 1950s and were used mainly for oxygen monitoring. Miniaturized sensors came in the mid-1980s for the detection of many different toxic gases with a good sensitivity and selectivity. The physical size, geometry, selection of various components, and the construction depends on the intended application and use. Different electrochemical sensors may appear very similar but are constructed with different materials such as sensing electrodes, electrolytic composition, and porosity of hydrophobic barriers. All components of the sensor play a crucial role in determining the overall performance of the sensors. At present, a wide variety of electrochemical sensors are being used for many applications including portable gas-detection systems. Sensing gases include highly poisonous chlorine (Cl_2), arsine (AsH_3), germane (GeH_4), and phosphine (PH_3).

These types of devices to detect CO, NO_x, H_2, and different hydrocarbons are also reported in the open literature. The prototype devices display many desirable characteristics in terms of size, cost, and sensitivity for many applications, particularly for combustion monitoring, but many do not meet the required long-term stability [26]. A typical representation of a mixed potential device is shown in Figure 4.4 where a solid-state cell of Electrode-1/Solid-electrolyte/Electrode-2 is constructed. Both electrodes are exposed to the analyte gas, usually a mixture containing oxygen and an oxidizable or reducible gas. Mixed potentials of different voltages develop on each electrode due to differences in electrokinetic redox rates of these electrodes. The

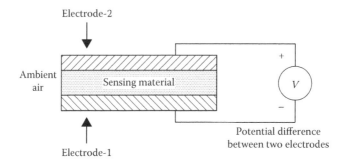

FIGURE 4.4
Schematic of a mixed potential sensor fabricated from yttria-stabilized zirconia. Both the electrodes are exposed to a gas mixture containing an oxidizable gas or reducible gas and O_2. Potential difference is measured across two electrodes.

device response voltage is the difference in mixed potential values attained by each of the electrode. Figure 4.4 is an example of one such mixed potential device fabricated using gold and platinum electrodes deposited on zirconia surfaces. Electrode materials, geometry, morphology, and nature of solid electrolyte will influence the mixed potential response toward the test gases. The substitution of refractory oxide electrodes in place of metal electrodes offers a promising and improved performance in the selectivity and long-term stability of these mixed potential sensors.

4.2 Sensor Properties

Each sensor exhibits certain properties and is measured by different techniques. Many researches define this parameter differently. To measure sensitivity of CO_2 sensing layers, Keller et al. [27] defined the sensor response, S, as $S = R_{CO_2}/R_{air}$, where R_{CO_2} and R_{air} are the electrical resistances of the sensor device measured at synthetic air (80% N_2, 20% O_2) and synthetic air with additional concentrations of CO_2, respectively. Cantalini et al. [28] defined the response (S), as the ratio between resistance R_g in gas and R_a in air, that is, $S = R_g/R_a$. Sberveglieri et al. [29] defined this as the relative change in the conductance $R = \Delta G/G_{air}$ for air reducing gases.

4.2.1 Sensitivity of Gas Sensors

Sensitivity, S, is defined as the ratio of change of resistance in test gas $\Delta R = R_a - R_g$, to the value of resistance in air R_a where R_g is the sensor resistance in the presence of the test gas [16]:

$$S = \frac{\Delta R}{R_a} = \frac{|R_a - R_g|}{R_a} \tag{4.7}$$

Sensitivity is also defined by the formula $S = (R_0 - R_{gs})/R_{gs}$, where R_0 is the resistance of the sensor before passing gas and R_{gs} is after passing gas and reaching the saturation value [30]. Others have defined this by $S = R_a/R_g$, where R_a and R_g express the resistance of the sensor in air and in detecting gas [31–33].

The sensitivity of the semiconducting oxide gas-sensitive sensor can usually be empirically represented as $S_g = Ap_g^\beta$, where p_g is the target gas partial pressure, which is in direct proportion to its concentration, and the sensitivity is characterized by the prefactor A and exponent β [34]. β may have some rational fraction value (usually 1 or ½), depending on the charge of the surface species and the stoichiometry of the elementary reactions on the surface.

4.2.2 Concentration-Dependent Sensitivity

At low test-gas concentration values, the sensitivity values increase steeply with an increase in the concentration values; however, at higher concentrations, the increase in the gas-sensitivity values becomes more gradual in nature. With a fixed surface area for each sensing element, a lower gas concentration implies a lower coverage of gas molecules on the available sensor surface, and hence, lower surface reactions occurred. An increase in the gas concentration raises the surface reactions due to a larger surface coverage. Further increase in surface reactions will be gradual when the saturation point on the coverage of molecules at the fixed surface area is reached. Further increase in the concentrations may not yield much sensitivity due to nonavailability of fresh surface area and it may even saturate at a fixed value as no further place is available for surface reactions to take place. This ultimately limits the signal available for analysis.

4.2.3 Temperature-Dependent Sensitivity

The sensitivity also depends on the temperature of the sensing element. Most of the metal-oxide sensors operate at a temperature above the room temperature range extending up to several hundred degrees centigrade and it is necessary to know the correct operating temperatures of these sensing elements. To obtain good sensitivity, one should know and operate these elements accordingly. Depending on the material and the detecting gaseous species, the values differ. In a typical case, the sensitivity increases gradually with temperature and becomes less gradual at higher temperature values. From linear dependency, it deviates to a maximum value and beyond this point the sensitivity falls rapidly. This type of behavior is being exhibited by almost all the metal-oxide-based sensors. Figure 4.5 shows a sensitivity graph illustrating the response of SnO_2 sensor to 100 ppm methane gas in dry air as a function of sensor operating temperature [35]. The sensitivity

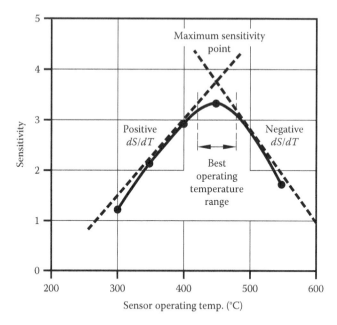

FIGURE 4.5
Response of SnO_2 sensor to 100 ppm of CH_4 gaseous species in dry air as a function of sensor operating temperature. Sensor sensitivity with temperature changes from positive value to negative value. Best operating range is indicated here for sensor operation. (Experimental data points are from *Prog. Solid State Chem.*, 33, Niederberger, M., Garnweitner, G., Pinna, N., and Neri, G., Non-aqueous routes to crystalline metal oxide nanoparticles: Formation mechanisms and applications, 59–70, Copyright 2005, with permission from Elsevier.)

increases continuously from 300°C up to 450°C and this is the best performance point. At higher temperatures, there is a fall in the sensitivity. Sensor sensitivity with temperature changes from positive value to negative value. Further rise in temperature will reduce the sensitivity values. From the experimental results, one can identify the best operating temperature range for the sensing element.

Similar characteristics are seen for the sensors based on In_2O_3 sensor for detecting 2 ppm NO_2 gas, as shown in Figure 4.6 [35]. Here also, the sensitivity initially rises with temperature and after reaching a maximum value it falls steadily moving toward a saturation value. The sensor sensitivity with temperature changes from positive value to negative value. Best operating range is indicated here for sensor operation. Positive slope indicates the reaction limited case whereas the negative slope is due to binding energy of the gaseous species.

Creation of high-temperature zones at small defined areas is crucial for operating small and micro-gas-sensing devices. Simple silicon surfaces take the temperature range of 50°C–150°C when directly heated. Its good thermal

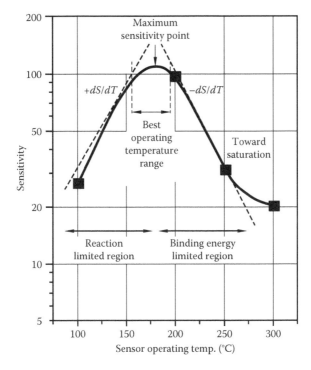

FIGURE 4.6
Response of In_2O_3 sensor to 2 ppm of NO_2 as a function of sensor operating temperature. Sensor sensitivity with temperature changes from positive value to negative value. Best operating range is indicated here for sensor operation. Positive slope indicates the reaction limited case, whereas the negative slope is due to binding energy of the gaseous species. (Experimental data points are from *Prog. Solid State Chem.*, 33, Niederberger, M., Garnweitner, G., Pinna, N., and Neri, G., Non-aqueous routes to crystalline metal oxide nanoparticles: Formation mechanisms and applications, 59–70, Copyright 2005, with permission from Elsevier.)

conducting properties limit the temperature rise. In case of silicon-on-insulator structures one can achieve a temperature range between 50°C and 350°C due to the presence of an insulator. Micromachined silicon structures, such as suspended platforms, defined using MEMS bulk and surface micromachining technique, are best suited for high-temperature creation in small localized areas. The air insulation provides the required thermal isolation. This way, one can create a temperature between 50°C and 500°C. Beyond 500°C, it is better to think of hybrid microstructures with external heating facility and keep the sensor away from the remaining sensor electronics. One has to design by carefully selecting the material, its operating temperature range, and the suitable technique to use these material compositions for sensor applications.

It is reported that by properly selecting the microhotplates, it is easy to achieve a temperature of the order of 350°C where most of the metal oxides show very good response properties. Such fabrication techniques are compatible with standard industrial CMOS technology as well [36]. The hotplate gives excellent temperature homogeneity with fluctuations of less than 2% of the respective operation temperature on the heated area. The membrane power efficiency is reported to be 4.8°C/mW. The system also exhibits excellent thermal isolation between heated membrane area and the circuitry area on the bulk chip.

4.2.4 Response and Recovery Times

For sensor applications, it is also very important for a sensor to react accurately to quick changes in gas concentration values. Optimization of gas sensor to respond quickly to sensing requirements is not free from its problems. Many gas sensors exhibit slower response times. Here, response time (τ_{90}) is typically defined as the time it takes for a sensor to reach a 90% of full-scale reading after being exposed to a full-scale concentration of a given gas. In certain cases, τ_{63} or τ_{80} are also used. In the majority of cases, it is τ_{90} criteria adopted for the evaluation of this time constant. Similarly, the recovery times are also important. These two time constants generally differ and in some cases recovery may pose a problem. In such cases, it is better to increase the temperature of the sensor above 50°C of its operating temperature, a process known as "regeneration of sensor," before it is reused. This regeneration cycle may be repeated periodically for the better efficiency of the sensing element.

The response time is proportional to particle size, d, of sensing material when the sensor kinetics is controlled by surface reaction and that is proportional to d^2 when the kinetics is controlled by diffusion. Regardless of the rate-limiting step, the response time decreases with decreasing particle size [3]. Therefore, it is expected that response time is much shorter if the sensing material is a nanosized particle.

5

Advantages of Nanomaterials

A vast amount of true multidisciplinary fundamental knowledge is currently being explored and linked with different fields of science and technology. It will lead to a tremendous amount of new in-depth understanding as well as to the fabrication of novel high technological devices in many fields of applications from electronics to medicine. Recent advances in synthesis, processing, and microanalysis are enabling the routine production of well-characterized materials with structure that varies on the length scale of several nanometers. Dealing with such nanomaterials is one such new concept to bring out new avenues in the field of gas sensors. Room-temperature hydrogen sensor [37] based on palladium nanowires is one such example.

Nanostructures, the structures that are defined as having at least one dimension between 1 and 100 nm, have received steadily growing interests as a result of their peculiar and fascinating physical and chemical properties and the applications superior to their bulk counterparts. The ability to generate such minuscule structures is essential to much of modern science and technology. There are a large number of opportunities that might be realized by making new types of nanostructures or simply by downsizing existing microstructures into the 1–100 nm regimes. The most successful example is being provided by silicon microelectronics fabrication technology, where "smaller" has meant greater performance ever since the invention of integrated circuits: More components per chip, faster operation, lower cost, and less power consumption are major achievements of fabricated devices, which have gone much beyond the imagination of many intellectuals of the scientific community. New examples are semiconductor quantum dots and superlattices, polymer nanocomposites, multilayer coatings, combination of microelectronic and optoelectronic devices, and the latest issues relating to microelectromechanical sensors.

Nanosized materials have been widely used to produce new semiconductor gas sensors. These materials have received considerable interest in recent years because of their unique properties both in physics and chemistry. It is now well established that these materials in this form exhibit some interesting properties. There are also multiple advantages of nanostructured conducting materials as compared to the coarse ones in their applications. The nanostructured materials can effectively increase the surface area for catalytic reaction and electrode reaction and result in an increase in the catalytic activity for oxidation to take place [38]. These materials also greatly reduce the necessary sintering temperature for densification, which is beneficial for many scientific applications. The sensitivity as well as the range of dynamic response of a gas sensor can be improved significantly when nanotechnology methods are used. For nanocrystalline semiconductors, nearly all conduction electrons are trapped in the surface states, which will enhance the sensor response for surface reactions. High porosity and large surface areas are desirable for numerous applications of nanostructures to fulfill the demand of high efficiency and high surface activity.

Nanocrystalline materials can be classified into different categories depending on the number of dimensions that are nanosized, with dimensions typically lower than 100 nm. These materials have emerged as attractive alternatives to conventional materials by virtue of their prominent electronic and chemical properties. Up to now, semiconducting metal-oxide sensors have been widely investigated due to their small dimensions, low cost, and high compatibility with microelectronic processing. In the case of the polycrystalline (ceramic or film) devices, only a small fraction of the species adsorbed near the grain boundaries is active in modifying the electrical transport properties. This causes the low maximum sensitivity because of the limited surface-to-volume ratio, which is difficult to overcome in normal materials. Owing to the great surface activity provided by their enormous surface areas, nanomaterials are expected to exhibit higher gas sensitivity. Adsorption of electron-acceptor gaseous species leads to band bending and the formation of a surface depletion layer due to capture of the free charge carriers at the surface. An important feature of these nanodimensional materials is that they bridge the crucial dimensional gap between the atomic and molecular scale of fundamental sciences and the microstructural scale.

In view of the aforementioned technical point, nanostructures of metal-oxide materials are of considerable interest in the area of chemical sensing due to the improved sensitivity of these structures toward the gaseous species when compared with the polycrystalline materials. The improved sensitivity of the nanostructured oxides is due to the availability of high surface area for each grain boundary site when compared with the polycrystalline materials. Furthermore, the higher gas sensitivities of nanostructures may make it possible to develop pure oxide sensing films that do not require additional catalytic additives for attaining acceptable performances. Chemically sensitive nanostructures integrated onto microsystems can pave the way for making portable chemical sensors, for monitoring toxic gases, and for field applications. The first evidence that oxide nanoparticles improve sensitivity to reducing gases was obtained by Ogawa et al. [39]. The Hall measurement studies clearly indicated that the ultrafine particle films are highly sensitive to reducing gases.

The possible nanostructures reported in the literature have features such as nanobelt, nanocage, nanocomb, nanocuboid, nanocylinder, nanodot, nanofiber, nanofilm, nanogranule, nanohollow particle, nanolayer, nanoneedle, nanoparticle, nanoplatelet, nanoporous structure, nanopowder, nanoring, nanorod, nanosaw, nanosheet, nanotransistor, nanotube and tubule, and nanowires. It is not possible that every metal oxide exhibits all the mentioned nanofeatures but most of the structures are possible by using these metal oxides for gas-sensing applications. Details about these nanostructures, regarding metal oxides, and their syntheses are explained in Chapter 6 of this book and are listed in Table A.2.

Nanotechnology, which exploits materials of dimensions smaller than 100 nm, is addressing the challenge and offering exciting new possibilities. In addition to a large surface-to-volume ratio and a Debye length comparable to the small size, they demonstrate different but superior sensitivity to surface chemical processes. For several years, many research works and commercial products have sought to exploit the properties of these nanocrystals as a constitutive material in the sensitive layer for gas-sensing applications. Of particular interest for metal-oxide-based conductance sensors are the grain sizes below the Debye length of electrons. Charge-transfer reactions during chemisorption and catalytic reactions at surfaces lead to uniform Fermi energy shifts instead of band-bending effects. Such nanostructured semiconductor oxides offer some advantages: a larger surface-to-volume ratio and a grain size of few nanometers, which is comparable to the depth of the space charge layer that surrounds the nanograin [40]. These specific features make nanocrystalline semiconducting oxides very promising for development of high-sensitivity, fast-response gas sensors, in which just surface processes play the key role in the formation of a sensor signal.

A variety of inorganic nanomaterials, including single element and compound semiconductors, have been successfully synthesized till date. With their in-depth physical property characterizations, they have been demonstrated to be promising candidates for future nanoscale electronic, optoelectronic, and sensing-device applications. The best sensors should be based on materials whose equilibrium is limited by the surface-exchange reaction rather than bulk diffusion and this is possible only with nanomaterials. Among the semiconductors, metal oxides especially stand out as one of the most versatile materials, owing to their diverse properties and functionalities that they exhibit.

This concept and synthetic method allows the design and the creation of metal-oxide nanomaterials with novel morphology, texture, and orientation and enables one to probe, tune, and optimize their properties. The combinations of such a variety of distinctive properties and applications with the unique effects of low dimensionality at nanoscale make the development of metal-oxide nanostructures an important challenge from both fundamental and industrial standpoints. It is well known that in sensors made out of nanomaterials instead of micromaterials, the responding rate and sensitivity is improved and the working temperature will be reduced due to the large surface area available. With decreasing grain size, the contribution from the surface energy to the total free energy of the system increases dramatically, leading to high activity within these material interactions. In addition, a catalyst with smaller particle size will have better catalytic activity.

Nanomaterials, including nanotubes, nanowires, nanobelts, and nanodots, usually exhibit a wide range of electrical and optical properties that depend on both size and shape. They are of interest in the theoretical understanding of quantum effects and have great potential application in many branches of science and technology. Control of the shape of these materials has remained an interesting and important topic in the field of nanomaterial synthesis. However, the challenge of controlling the shape of nanoparticles has met with limited success so far and efforts are still in progress. With a large surface-to-volume ratio and a Debye length comparable to the nanowire radius, the electronic property of the nanowire is strongly influenced by surface processes, yielding superior sensitivity than its thin-film counterpart. Fortunately, metal-oxide-nanowire-based gas sensors have demonstrated significantly higher sensitivity even at room temperature.

Every sensor device possesses a sensing body (or resistor), which is a porous assembly of tiny grains of metal oxide. Under exposure to air, oxygen is adsorbed on the grains as anionic species (typically O^-), inducing an electron-depleted layer to increase surface potential and work function [41]. No charge carriers are present in these depleted regions. In cases where the grains are connected to their neighbors through the grain boundaries, a potential barrier for migration of electrons, often called double-Schottky barrier, is formed across each grain boundary, as shown in Figure 5.1a. This barrier plays a dominant role in determining the resistance of the sensing body. When a reducing gas is brought into contact, the oxygen adsorbates are consumed, leading to a mitigation in potential barrier and then also in R. Thus, R decreases with increasing partial pressure (p) of the reducing gas. On the other hand, on contact with an oxidizing gas like NO_2 in air, target gas and oxygen are adsorbed competitively as anionic ions on the same grains. If adsorption of the target gas is stronger than that of oxygen, the surface potential barrier will increase upon exposure to the target gas, resulting in an increase in R. Electrons can also be transported by tunneling through a small gap between oxide grains even when grain boundaries are not formed effectively throughout the sensing body, as shown in Figure 5.1b. Even in this case, tunneling current is attenuated by the surface potential of oxide grains, resulting in almost the same dependence of R on p. Here, the vacuum level is indicated for reference.

For simplicity, if we assume that the radii of oxide grains (n-type semiconductor) are much larger than Debye length. This allows us to approximate a depletion layer of each grain by a one-dimensional (1-D) model, as shown in Figure 5.2, where x denotes depth from the surface. When a certain surface state is present at an energy level below the Fermilevel under the flat band condition, electrons are transferred from the conduction band to the surface state and this transfer continues until the Fermilevel becomes constant throughout, from the surface to bulk in equilibrium state [41]. The resulting depletion layer is well described by a depth profile of potential energy of electrons, $qV(x)$, and well characterized by surface potential barrier, qV_s, and depletion depth,

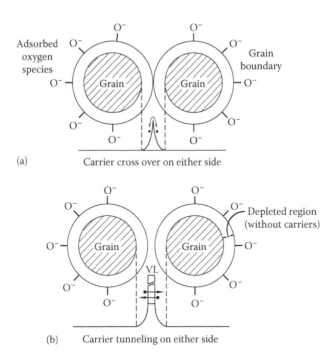

FIGURE 5.1
Models of electron transfer between grains: (a) double-Schottky barrier model and (b) tunneling model. VL stands for vacuum level. (Reprinted from *Sens. Actuators*, B128, Yamazoe, N. and Shimanoe, K., Theory of power laws for semiconductor gas sensors, 566–573, Copyright 2008, with permission from Elsevier Science B.V.)

w, as shown in the figure. This depletion region width is a moving boundary within the grain region. Figure 5.2 shows the moving boundary within the grain region altering depletion region width depending on the surface reactions. Here, q is represented as elemental charge of electron and $V(x)$ and V_s are electric potential at given depth and surface, respectively.

In a low-grain-size metal oxide almost all the carriers are trapped in surface states and only a few thermally activated charge carriers are available for electrical conduction. In this configuration, the transition from "activated" to strongly "nonactivated" carrier density, produced by target gas species, has a great effect on sensor electrical conductance. The challenge became to prepare materials with small crystalline size that were physically stable when operated at high temperature for longer periods. The strong dependence of the grain–grain conductivity from the band bending, as shown in Figure 5.3, is the main reason for the high sensitivity of SnO_2-based gas sensors to reducing gases. The polycrystalline structure could be connected by electrodes in order to measure the gas-depending sensor resistance that results from the network structure and the resistance of the several grains contacts [20,42] as shown in the figure. In the physical model, polycrystalline structure of the

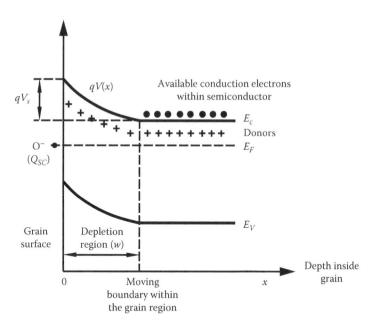

FIGURE 5.2
Schematic drawing of surface-depletion layer formed by oxygen adsorption for an *n*-type semiconductor. Diagram shows the moving boundary within the grain region altering depletion region width. (Reprinted from *Sens. Actuators*, B128, Yamazoe, N. and Shimanoe, K., Theory of power laws for semiconductor gas sensors, 566–573, Copyright 2008, with permission from Elsevier Science B.V.)

sensor surface shows the adsorbed oxygen species and the depletion regions. In the band model, conduction band electrons and the tunneling of charge carriers across the barriers are shown.

Sensor performance depends on percolation path of electrons through intergranular regions, by varying small details in the preparation process. Both thin and thick film electrical properties drift due to the grain coalescence, porosity modification, and grain-boundary alteration [43]. These effects become more critical because the metal-oxide layers must be kept at a relatively high temperature in order to guarantee the reversibility of chemical reactions at surface. Thus, several solutions have been put forward to stabilize the nanostructures. Addition of a foreign element or phase to the material is one such effort and technique generally adopted to maintain the physical structure.

The peculiar characteristics and size effects of nanosized structures make them interesting for both fundamental studies and potential nanodevice applications, leading to a third generation of metal-oxide gas sensors. Their nanosized dimensions generate properties that are significantly different from their coarse-grained polycrystalline counterpart. Surface effects appear because of the magnification in the specific surface of nanostructures,

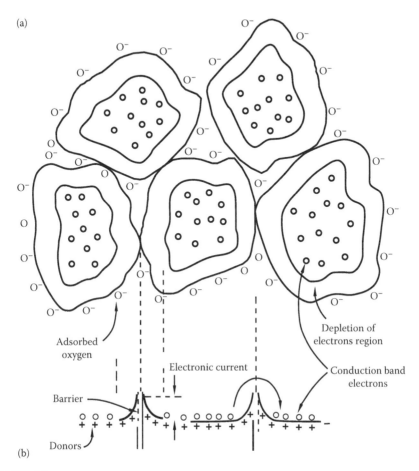

FIGURE 5.3
Polycrystalline structure of the sensor surface in (a) physical model and (b) the corresponding band model. Diagram shows the adsorbed oxygen species and the depletion regions within the grain regions. (Adapted from original papers Madou, M.J. and Morrison, S.R., *Chemical Sensing with Solid State Devices*, Academic Press, London, U.K., Copyright 1989 Academic Press; reprinted from *Phys. Stat. Mech. Appl.*, Bläser, G., Rühl, Th., Diehl, C., Ulrich, M., and Kohl, D., 266, Nanostructured semiconductor gas sensors to overcome sensitivity limitations due to percolation effects, 218–223, Copyright 1999, with permission from Elsevier Science B.V.)

leading to an enhancement of the properties related to that, such as catalytic activity or surface adsorption processes. These specific properties are the basic phenomenon underlying the solid-state gas sensors.

In current material science research, the use of nanosized materials for gas sensors is rapidly arousing interest in the scientific community. One reason is the surface-to-bulk ratio and another is that the conduction type of the material is determined by the grain size of the material. When the grain size is small enough (the actual grain size D is less than two times the space-charge depth L), the material resistivity is determined by grain control, and the material's conduction type becomes surface conduction type [44]. Hence, the grain-size reduction becomes one of the main factors in enhancing the gas-sensing properties of semiconducting oxides.

The equilibrium phase of nanosized particles may deviate from that of the bulk owing to the contributions of the surface energy and surface stress. Knowledge and understanding of the behavior of small particles and inclusions is of importance for many engineering materials, particularly with the ongoing trends toward materials with a controlled nanostructure. In this context, the study of phase transitions of small particles and inclusions deserves special attention [45], since it will probably play a key role in the improvement of the understanding and control of the microstructure.

Nanotechnology could provide sensing materials capable of improving the sensitivity, selectivity, stability, and the speed of sensor technology. One recent area of interest involves applying quasi-1-D metal-oxide nanostructures, such as nanowires, nanotubes, nanobelts, etc., as prospective gas-sensing elements [46]. The nanostructures maintain high surface area to bulk volume ratio and the comparability of their radii and their corresponding Debye lengths allow rapid transduction of surface interactions into measurable conductance variations. It also opened a totally new type of electronic device where nanowire-based active device is being proposed. The best example for one such proposal is an open-gate FET transistor.

To explain the effect of grain size on the sensitivity of metal-oxide gas sensors, a semiquantitative model was proposed by Rothschild and Komem [47], which is shown in Figure 5.4. According to this model, the sensing material consists of partially sintered crystallites connected to their neighbors by narrow necks. Those interconnected grains form larger aggregates and are connected to neighbors through the grain boundaries. Three different cases can be distinguished according to the relationship between the grain size D and the width of the depletion layer thickness L that is produced around the surface of the crystallites due to chemisorbed adions. When $D \gg 2L$ most of the volume of crystallites is unaffected by surface interactions with the gas phase. In this case, the predominant effect of the ambient gas atmosphere on the sensor conductivity is introduced via the GB barriers for intercrystallite charge transport from one grain (agglomerate) to another.

The electrical conductivity σ depends exponentially on the barrier height $q|V_B|$: $\sigma \propto \exp(-q|V_B|/kT)$, where q is the elementary electron charge, V_B is the GB potential, k is Boltzmann constant, and T is temperature. According to depletion approximation $q|V_B| \propto (N_t^-)^2$, where N_t^- is the trapped charge density at the surface of the crystallites. N_t^- can be modified by charge-transfer interactions with reactive gases such as O_2, NO_x, and CO, and the conductivities are sensitive to the ambient gas composition. Thus, for large grains ($D \gg 2L$) the

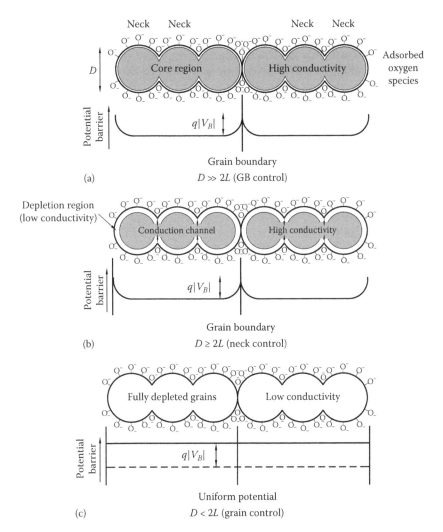

FIGURE 5.4
Schematic model of the effect of the crystallite size on the sensitivity of metal-oxide gas sensors: (a) $D \gg 2L$ (grain-boundary controlled); (b) $D \geq 2L$ (neck controlled); and (c) $D < 2L$ (grain controlled). (Reprinted with permission from Rothschild, A. and Komem, Y., The effect of grain size on the sensitivity of nanocrystalline metal-oxide gas sensors, *J. Appl. Phys.*, 95, 6374–6380, Copyright 2004, American Institute of Physics.)

gas-sensing mechanism is controlled by the GB barriers. Furthermore, the GB barriers are independent of the grain size and therefore the sensitivity is independent of D value. The details are shown in Figure 5.4a.

As the grain size decreases, the depletion region extends deeper into the grains and consequently the core region, which is relatively conductive with respect to the depletion region adjacent to the surface, becomes smaller. When D approaches but still larger than $2L$, that is, when $D \geq 2L$, the depletion region that surrounds each neck forms a constricted conduction channel within each aggregate, as depicted in Figure 5.4b. Consequently, the conductivity depends not only on the GB barriers but also on the cross-sectional area of those channels. This area is proportional to $(X - L)^2$, where X is the neck diameter, which is proportional to the grain size $D(X \approx 0.8D$ according to Xu et al. [48]). As a result, the conductivity is a function of the ratio X/L (or X/D). Since $L \propto N_t^-$ and N_t^- is modulated by the surface interactions with the gas phase, the effective cross-sectional area of the current path though the grains is sensitive to the ambient gas composition. The current constriction effect adds up to the effect of the GB barriers, and therefore the gas sensitivity is enhanced with respect to the former case (as shown $D \gg 2L$, cf. Figure 5.4a). Furthermore, the sensitivity to gases becomes grain-size dependent and it increases when D decreases. Details of neck-controlled sensitivity are shown in Figure 5.4b.

When $D < 2L$ the depletion region extends throughout the whole grain and the crystallites are almost fully depleted of mobile charge carriers, as shown in Figure 5.4c. As a result, the conductivity decreases steeply since the conduction channels between the grains are vanished. The energy bands are nearly flat throughout the whole structure of the interconnected grains, and since there are no significant barriers for intercrystallite charge transport, the conductivity is essentially controlled by the intracrystallite conductivity (grain controlled). It was found empirically that the highest gas sensitivity is obtained in this case, that is, when $D < 2L$.

Since the electrical conductivity of the films is very crucial and systematic measurement of this important parameter reveals the internal grain property of the films under study. In this connection, a new conductivity sensor structure is defined by Hoel et al. [49] to measure the conductivity change in nanoparticle thin films, as shown in Figure 5.5. Figure 5.5a and b shows two different arrangements for testing the sensor properties of WO_3 nanoparticle

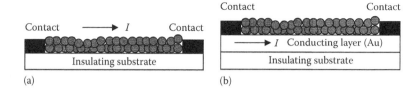

FIGURE 5.5
Sensor constructions based on nanoparticles. (a) Shows a "conventional" sensor, detecting conductivity change of the nanoparticle film as current I flows through it; (b) illustrates an arrangement to measure conductivity fluctuations of a thin conducting (gold) layer covered by nanoparticles. (Reprinted from *Curr. Appl. Phys.*, 4, Hoel, A., Reyes, L.F., Heszler, P., Lantto, V., and Granqvist, C.G., Nanomaterials for environmental applications: Novel WO_3-based gas sensors made by advanced gas deposition, 547–553, Copyright 2004; *Smart Mater. Struct.*, 11, Hoel, A., Ederth, J., Kopniczky, J., Heszler, P., Kish, L.B., Olsson, E., and Granqvist, C.G., Conduction invasion noise in nanoparticle WO_3/Au thin-film devices for gas sensing application, 640–644, Copyright 2002, with permission from Elsevier B.V.)

films. A "conventional" arrangement is shown in Figure 5.5a, where current passes though the nanoparticle film and the mean resistance change is measured as the layer is exposed to the probing gas. Figure 5.5b illustrates a recently suggested new type of arrangement for nanostructured sensing films [50] for the same purpose. In this case, current is driven through a thin conducting layer, such as gold in the present case, covered by sensor-type nanoparticles. Resistivity fluctuations of the conducting layer are measured as the device is exposed to the test gas. Sensitivity and other parameters are to be studied carefully in this case and this approach is expected to provide better information on nanoparticles.

Although the reduction in the size of the sensing element and/or the transducer in a sensor are important, in order to miniaturize the devices, there are many issues in nanoscience one has to deal with for the nanosensor device development. New effects appear and play an important role that is often related to quantum mechanics and quantum mechanisms. Consequently, important characteristics and quality parameters of the nanosensors can be improved over the case of classically modeled systems by merely reducing the size. For example, sensitivity can increase due to better conduction properties [51]. The limits of detection can be lowered further and very small quantities of samples are possible to analyze. Direct detection is also possible, in this case, without using any labels, and some reagents can be eliminated.

Nanobelts of semiconducting oxide, with a rectangular cross section in a ribbon-like morphology, are very promising for gas-sensing application due to the fact that the surface-to-volume ratio is very high, the oxide is single crystalline, the faces exposed to the gaseous environment are always the same and the size is likely to produce a complete depletion of carriers inside the belt structures. Not only is the reported deposition technique very simple and cheap, but the size and shape can be easily controlled in this type of structures [19]. In the polycrystalline and thick-film devices, only a small fraction of the species adsorbed near the grain boundaries is active in modifying the device electrical transport properties. In the new sensors, based on single crystalline nanobelts as mentioned, almost all of the adsorbed species are active in producing a surface depletion layer, an essential condition for gas sensors as shown in Figure 5.5.

Nanoparticles are clusters of a few hundred to a few thousand atoms that are several nanometers long. Because of their size, which is of the same order as the de Broglie wavelength associated with the valence electrons, following the wave–corpuscle duality principle, with a wavelength λ, nanoparticles behave electronically as zero-dimensional quantum dots with discrete energy levels that can be tuned in a controlled way by synthesizing nanoparticles of different diameters. A quantum dot is a location that can contain a single electrical charge, that is, a single electron. The presence or absence of an electron changes the properties of a quantum dot in some useful way and they can, therefore, be used for several purposes such as information storage or useful transducers in sensors [51]. Nanoparticles have outstanding size-dependent optical properties that have been used to build optical nano-sensors primarily based on noble-metal nanoparticles or semiconductor quantum dots.

The 1-D nanostructures such as nanowires, nanorods, nanobelts, and nano-tubes have become the focus of intensive research activity owing to their unique applications in different fields. In particular, nanosized metal-oxide particles also show a quantum size effect, which causes peculiar optical and magnetic characters. Metal-oxide nanoparticles are important materials; they have been studied with regard to the optical, electronic, and chemi-cal properties of quantum dots and are widely used in applications, such as catalysis, electrochromic windows, sensors, and magnetic materials, due to the extremely small size of the nanocrystals and their surface chemistry [52]. From bulk to thin film and from thin film to nanoscale dimensions strain effects also alter effective mass parameter values [53]. Not much information is available on these issues for metal-oxide structures. In this book, we con-centrate mainly on the gas-sensing applications.

6

Synthesis and Characterization of Nanometal Oxides

A variety of methods have been utilized to grow nanostructures. According to the synthesis environment, they can be mainly divided into two categories: vapor phase growth and liquid (solution) phase growth. Most of the metal oxide nanostructures are grown via the well-developed vapor phase technique, which is based on the reaction between metal vapor and oxygen gas. The governing mechanisms are the vapor–liquid–solid (VLS) process and vapor–solid (VS) process. On the other hand, solution-phase growth provides more flexible synthesis process and an alternative to achieve lower cost. Table A.2 shows broadly the different nanostructures obtained through different routes using a wide variety of techniques to obtain these metal-oxide nanostructures. They are arranged in alphabetical order, for simplicity, and the complex (or mixed) oxides are kept at the end. Listed are the oxides of aluminum, bismuth, cadmium, cerium, cobalt, copper, iron, gallium, indium, iridium, molybdenum, niobium, nickel, tin, tellurium, titanium, tungsten, zinc, zirconium, and mixed oxides. Although, a large variety of synthesis approaches have been reported, the preparation of crystalline metal oxides on the nanoscale range is still a big challenge.

To achieve smaller grain size requires a close control over the other characteristics of nanocrystals. If such control is not maintained, then sensor properties such as stability, sensitivity, and sensor response to different gases is not the same from batch to batch or from laboratory to laboratory [207]. Moreover, some of these processes require high energetic budgets and are time consuming, aspects that make them unsuitable for the mass production of these gas sensors.

High-energy ball milling technique was employed by Tan et al. [44] to synthesize various nanostructured metal-oxide semiconductor materials with average particle size down to several nanometers. The mechanical alloying processes for both ethanol sensing materials and oxygen sensing materials have been investigated and found to be as good as those prepared by other techniques. Nanosized α-Fe_2O_3-based solid solutions mixed with different mole percentages of SnO_2, ZrO_2, and TiO_2 have been obtained separately and fabricated into thick-film sensor devices for gas-sensing applications. The results have shown that the high-energy ball milling is a very effective route to prepare nanostructured solid solutions at room temperature for gas-sensing applications.

Sol–gel chemistry allows the combination at the nanosize level of inorganic and organic or even bioactive components in a single hybrid composite, providing access to an immense new area of materials science. The nano-structure, the degree of organization, and the properties that can be obtained for such materials certainly depend on the chemical nature of their components, but they also rely on the synergy between them.

Oh et al. [66] suggested a new technique to prepare nanocrystalline structures of metal oxides. This is by applying ultrasonic spray pyrolysis technique and has several advantages such as simplicity and a low-cost alternative for the deposition of high-quality thin films. This method is also applicable for the drying of colloids, direct control of microstructure and size of nanoparticles, and good control of particle composition. When the precursor solution is atomized to form droplets and then pyrolyzed, the colloidal particles serve as seeds for nucleation leading to dense particle formation.

Transmission electron microscopy is the ideal tool and most productive method for analyzing the composition and structure of many fabricated nanostructures in two and three dimensions [208]. The surface chemistry is typically investigated by using Fourier transform infrared spectroscopy, which was proved to be a very relevant technique to obtain a thorough understanding of the surface phenomena at the origin of the gas-detection mechanism. This fundamental approach was considered as a critical step to refine the sensor optimization by tailoring both the surface chemical composition and reactivity of the nanoparticles during and eventually after their synthesis. It was also demonstrated that the systematic characterization of the nanoparticle surface at a chemical level guarantees the reproducibility of the sensor performance. This type of infrared analysis of semiconducting nanoparticles appeared to be extremely useful to get a relatively quick evaluation of the sensing potentiality of the materials as soon as they are synthesized. It also allows a fundamental study of the surface reactions at the origin of the gas-detection mechanism, thus making it possible to tailor the surface chemistry for maximizing the performance efficiency of these reactions responsible for the electrical conductivity variations in sensing element. By optimization of closed loops, it was possible to rapidly adjust the synthesis parameters for optimized sensing properties and to select the best batches of nanoparticles to be further processed.

Ordered mesoporous carbon was successfully used, by Kang et al. [209], as a rigid template for the formation of mesoporous non-silica materials. Various kinds of mesoporous inorganic materials such as alumina, zirconia, and titania can also be obtained from the carbon template by using the nanoreplication synthetic strategy. The method is especially suitable for preparation of the materials with highly crystalline frameworks. This noble nanoreplication technique for the preparation of mesoporous materials is believed to be a remarkable achievement in the field of porous materials. The results indicate that by this technique mesoporous metal oxides can be obtained. Relatively small pore size of the order of 3 nm is possible to achieve the metal-oxide nanostructures. There are many wide variety of techniques listed in Table A.2 for synthesizing different metal-oxide nanostructures.

7

Nanostructured Metal Oxides and Gas-Sensing Devices

Nanosized structures show outstanding properties in gas-sensor applications, especially at low operating temperatures, and, high sensitivity, particularly with lower grain sizes. Therefore, ultrafine layers or any feature that resembles lower feature sizes allow various improvements in their performance. The sensitivity of the sensor largely increases with the reduction of the feature size. To explain this property, we select a simple example. Lu et al. [107] showed that the nanosized SnO_2 material is sensitive to 500 ppm concentration of CO at near room temperatures, as shown in Figure 7.1. In this figure, the effect of nanoparticle size is compared to the sensitivity of CO gas. There is hardly any change in the sensitivity values observed for the particle size that is larger than 50 nm. Slow raising trend is noticed between particle sizes between 50 and 15 nm. When the particle sizes are smaller than 10 nm, the sensitivity shoots up to give a quick increase in the sensitivity values. This type of advantage is expected with almost all the metal-oxide nanostructures for gas-sensing applications.

Wide varieties of nanostructures for metal oxides are reported in open literature. Table A.2 lists these details and elaborates the different nanostructure shapes and the techniques adopted to synthesize them. They are arranged in an alphabetical order. The methods include aerosol; anodization in solution; auto combustion; cathodic vacuum arc method; chemical synthesis; condensation of vapor; controlled solid–vapor process; chemical vapor deposition (CVD) techniques such as atmospheric CVD, hot-wire CVD (HWCVD), plasma enhanced CVD, metal-organic CVD (MOCVD), and metal-seeding CVD; direct metal oxidation; electrospinning; focused high-intensity electron beam; hard template method; high-energy ball milling technique; hydrothermal technique; infrared irradiation; pulsed laser ablation; mechanical alloying; molecular beam epitaxy (MBE); pyrolysis techniques such as mist, spray, and ultrasonic; reactive gas evaporation technique; sol–gel technique; solution dip technique; solvothermal treatment; spin coating, spray drying, and spray precipitation; sputtering by magnetron and RF methods; supercritical fluid drying technique; supersaturated vapor nucleation process; temperature ramping process; using templates; thermal evaporation by heating method; vacuum arc deposition; and vapor phase deposition methods. To the field of gas-sensing applications, these features appear to be just a beginning. However, there is a lot of potential for the following nanostructures: nanobelts, nanobouquets, nanohelices, nanoribbons,

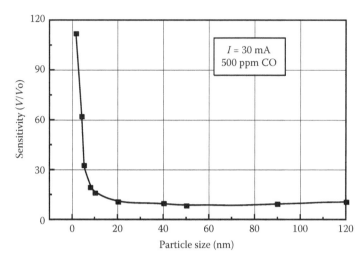

FIGURE 7.1
Effect of particle size on gas sensitivity of CO gas using nanosized SnO_2 material. (Reprinted from *Sens. Actuators*, B66, Lu, F., Liu, Y., Dong, M., and Wang, X., Nanosized tin oxide as the novel material with simultaneous detection towards CO, H_2 and CH_4, 225–227, Copyright 2000, with permission from Elsevier Science S.A.)

nanorods, nanotubes, and nanowires. As mentioned earlier, each metal oxide may not exhibit all the aforementioned features but several oxides do exhibit different nanofeatures depending on the process conditions. In this book, we plan to study these features one after another. These nanostructured metal oxides study are listed in an alphabetical order and there is no specific reason for this classification. This order is not to give any importance or any priority in their utility for gas-sensing applications. Though, many metal oxides are known to be sensitive to different gases, we mainly concentrate on reported nanostructured metal oxides, namely, oxides of aluminum, bismuth, cadmium, cerium, cobalt, copper, gallium, indium, iridium, iron, molybdenum, nickel, niobium, tellurium, tin, titanium, tungsten, vanadium, zinc, and zirconium. Mixed metal oxides are taken up separately at the end. Majority of these oxides are not in their stoichiometric concentration ratios but most of them are useful in sub-stoichiometric levels also. Some metal oxides show typical stoichiometric ratios not generally seen with other metal oxides. Tungsten oxides are the best example for such stoichiometric ratios and we shall take these up at a later stage.

7.1 Aluminum Oxides

Aluminum atom has an electronic configuration of $1s^2 \, 2s^2 \, 2p^6 \, 3s^2 \, 3p^1$ and has a valence of +3, in all compounds it form, with the exception of a few high-temperature monovalent and divalent formations. Since aluminum has only three electrons in its valence shell, it tends to be an electron acceptor. Its strong tendency to form an octet is shown by the tetrahedral compounds involving sp^3 hybridization. Aluminum ion has a coordination number 6 and forms compounds only at elevated temperatures. This metal does not form many organic compounds.

Sensor devices based on aluminum oxide materials, particularly Al_2O_3, are excellent sensing materials for humidity and they mainly operate in capacitor-type structures. These devices operate at near room temperature and show good stability as no surface chemical reactions take place when exposed. The Al_2O_3-based sensor is accepted as a volume effect device. The porous Al_2O_3 film is considered as a dielectric with pores whose conductivity varies as a function of moisture content trapped in these micropores. The moisture permeates through the top conductor layer of the sensing element and equilibrates on the pore wall. As a consequence, the admittance of the capacitive structure changes with humidity content. If these Al_2O_3 films are used as sensing elements at high humidity conditions, those exposed for sufficiently longer periods show sensor degradation in their sensitivity and drift in their parameter values are observed. Frequent regeneration of the sensing element is the best way to use these devices before the sensor degrades totally.

It is also reported that these aluminum oxide–based sensors detect H_2, CO_2, and O_2 when they are operated at higher temperatures, of the order of 450°C [210]. No experimental reports are available on the adsorption studies on Al_2O_3 surfaces. However, some theoretical calculations of gas adsorptions on (0001) surface of α-Al_2O_3 are available in the open literature. The hydroxylation of the α-Al_2O_3 (0001) surface has been considered theoretically by means of crystalline-orbital approach. The adsorption of a monolayer of –OH on top of the surface aluminum ions and hydrogen atoms bonded to the surface oxygen ions generates a new bond that is shifted a few eV above the top of the valence band in the bulk bandgap. The band is formed primarily from oxygen $2p$ and H $1s$ orbitals [211]. When compared with other humidity sensing materials, devices based on Al_2O_3 show exceptional quality with response times of the order of 1 min [212]. Though most of Al_2O_3 sensors are thin-film based, cylindrical-shaped sensors formed using alumina powder have also been reported [212].

Preparation of nano- or microscale hollow particle structures is of both scientific and technological interest. Particularly, the nano- or microscale hollow materials may lead to extensive applications in microvessel, catalytic support, adsorbents, light-weight structure materials, and thermal and electric insulators. Also, its hollow structure can be used as microencapsulates for drug-release control in pharmaceutical applications or gene gun bullet for gene transformation or gene therapy.

In order to obtain a regular solid boundary from the precursor agglomeration, Chou et al. [54] selected a highly polymeric precursor solution by applying a spray technique to atomize the precursor solution into nano- or micrometer scale droplets. Then the droplets were put into a precipitation agent. Under this condition, an isotropic contact surface is created and the agglomeration of precursor takes place when the droplets are surrounded by the precipitation agents including dehydration or neutralization agents. This method allows building nano- and micrometer-sized hollow sphere particles, in the case of precipitation taken place along a regular boundary, to form the rigid shell. After the precipitation, careful removal of the solvent within the formed rigid shell leaves hollow features. In this fabrication technique of microscale Al_2O_3 particle hollow structures, aluminum chlorohydrate (ACH) solution was selected as a precursor and acetone was used as a precipitation agent.

Figure 7.2 shows the hollow particles obtained by spraying the ACH solution into micro- or nanoscale droplets to get into contact with acetone.

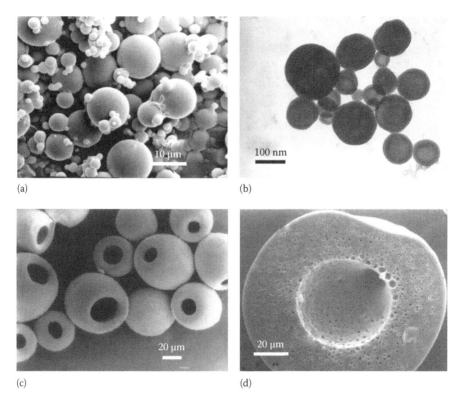

(a)   (b)

(c)   (d)

FIGURE 7.2
Scanning electron micrographs showing the morphology of nano- and microscale ACH particles: (a) a group of microsolid spheres prepared by conventional spray drying method; (b) transmission electron micrographs showing the morphology of nanoscale ACH sphere particles with hollow texture prepared by the spray precipitation method using precursor ACH solution and acetone; (c) a group of microbottle-shaped particles prepared by the spray precipitation method; and (d) cross section of a bottle-shaped particle of (c). (Reprinted from *Mater. Sci. Eng.*, A359, Chou, T.-C., Ling, T.-R., Yang, M.-C., and Liu, C.-C., Micro and nano scale metal oxide hollow particles produced by spray precipitation in a liquid-liquid system, 24–30, Copyright 2003, with permission from Elsevier B.V.)

Here, the ACH particles were obtained by conventional spray drying method exhibited solid sphere structure as shown in Figure 7.2a. By the spray precipitation method, the formation of solid-shell ACH particle has taken place on the surface of the sprayed droplet while it fell into acetone, water extraction environment. These spherical particles are in perfect shape with diameters ranging between 1 and 12 μm in this figure. Transmission electron microscope (TEM) micrographs indicated that these nano-ACH particles are hollow inside as shown in Figure 7.2b. The size of ACH droplets sprayed from a nebulizer could be controlled accurately and measured as about 20–70 μm in diameter with the shape of a bottle in which one side is open and the rest is closed. Figure 7.2c shows a group of such micro-ACH bottles formed and filtered out suitably. Most of the particles here are in the shape of an ellipsoid due to the pressure of fluid in the spraying process. The morphology of these particles is sensitive to the fluid motion in the liquid–liquid system and they get distorted easily into ellipsoid shape during the process of crystallization. A cross-sectional photograph of microbottle having hollow inside exhibits a clear porous, sponge shell structure with a well-defined annular shell thickness as shown in Figure 7.2d. The sponge pores are very small in size measuring in the nanometer range. The average diameter of the hole in the microbottle is about 30 μm and the shell thickness is about 5–20 μm. The inner diameter of this bottle is about 30 μm as shown in this figure.

Micro-Al_2O_3 hollow or bottle particles can also be synthesized by using $AlCl_3$ solution and hydrophilic–lipophilic precipitation agents such as tributylamine (TBA). This $AlCl_3$ precursor solution includes 15 g $AlCl_3 \cdot 6H_2O$ and 100 mL of water. As shown in Figure 7.3a, a microbottle enclosed with

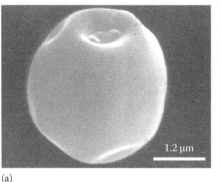
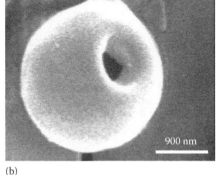

(a) (b)

FIGURE 7.3
SEM morphology of micro-aluminum-oxide hollow particle formed by spraying precipitation method with hydrophilic–lipophilic precipitation agents. (a) Micro-Al_2O_3 bottle prepared by using $AlCl_3$ solution–TBA system and (b) Al_2O_3 bottle obtained from burning out the TBA film. (Reprinted from *Mater. Sci. Eng.*, A359, Chou, T.-C., Ling, T.-R., Yang, M.-C., and Liu, C.-C., Micro and nano scale metal oxide hollow particles produced by spray precipitation in a liquid-liquid system, 24–30, Copyright 2003, with permission from Elsevier B.V.)

a gel-like film was synthesized after removal of TBA through vacuum drying process carried out at room temperature. After calcinations at 600°C for 4 h, the TBA gel film was burned off and micro-Al_2O_3 bottle shapes were formed, as shown in Figure 7.3b. As mentioned earlier, at high-momentum impact, a TBA film was coated on part of the particle surface, some part of $AlCl_3$ solution had been spouted out from this film. This type of structure is very attractive for gas-sensor applications as the opening allows gas interaction on both sides of the bottle-shaped structures. However, such studies are not yet reported for nanostructured Al_2O_3 films and for bottle-shaped structures. It is anticipated that such hollow structures may be useful for medical applications especially to release drugs.

In addition to the spray drying of droplet and the precipitation technique [54] for nanosphere particles, aluminum oxides are also deposited using cathodic vacuum arc deposition methods [55]. Nanowires of these oxides were also synthesized by using chemical etching of anodic alumina membrane [56]. By using anodized alumina templates, Kim et al. [213] have demonstrated the preparation of palladium nanowires. These nanowires were found to be excellent in detecting hydrogen gas with a quick response time of less than 20 s up to a concentration levels in the range of 0.2%–1% concentration. These anodized alumina templates are expected for many such applications of wider interest.

Recently Kocanda et al. [57] have demonstrated that by using nanoporous alumina films can be used for detection of cyclic organic compounds. They have demonstrated that nanoporous anodic aluminum oxide (AAO) materials exhibit a response to nonpolar organic molecules also. These compounds were structurally similar and also with relative sizes. The analytes used in this investigation were cyclohexane (C_6H_{12}), cyclohexene (C_6H_{10}), benzene (C_6H_6), methylbenzene (toluene, C_7H_8), 1,2-dimethylbenzene (*o*-xylene, C_7H_{10}), 1,3-dimethylbenzene (*m*-xylene, C_7H_{10}), and 1,4-dimethylbenzene (*p*-xylene, C_7H_{10}). Table A.3 shows the details of these sensing devices. This study of Kocanda et al. [57] has opened a new dimension for the nanoporous aluminum oxide as a sensor for cyclic volatile organic compounds (VOC).

7.2 Bismuth Oxides

Metal bismuth does not react with oxygen under normal conditions but forms an oxide coating at high temperatures. The yellowish-brown color of this oxide distinguishes this oxide from those formed by other metals.

Bismuth (III) oxide (Bi_2O_3) is a very important dielectric material and widely used in the field of gas-sensing applications and also as a solar-cell construction material. It is also useful as an oxide-ion conductor, as a piezo-optic material, and in transparent ceramic glass manufacturing processes. It is definitely a basic oxide but exists in three different structures: white rhombohedral, yellow rhombohedral, and gray-black cubical. Bismuth (II) oxide (BiO) is produced by heating the basic oxide.

In the pure form, Bi_2O_3 is a good smoke sensor at an elevated temperature in the range of 420°C–730°C. In the course of evaporation, bismuth oxide film loses oxygen and it is necessary to replenish the oxygen deficiency by heating the films in the oxygen environment. Preferred temperature range is determined by the presence of the metal bismuth in the film as established by film-specific electrical testing and x-ray measurements [214]. By using Sb_2O_3 as an activator, this Bi_2O_3 material is a good sensor for CO at 200°C [215]. The cross-sensitivity of these films with CH_4, liquid petroleum gas (LPG), and H_2 gases is substantial. It was felt that it is not the Bi_2O_3/SnO_2 interface that is taking part in CO detection but the compound $Bi_2Sn_2O_7$ phase is responsible. Further, it is also reported that the material is sensitive to H_2, CO_2, and O_2 at 450°C. Since the smoke mainly consists of CO and CO_2, this is probably a good smoke detector. Because of the high-temperature operation, the usage is not very promising for microsensors. Creation of such high temperatures of the order of 450°C is rather difficult.

Shen et al. [58] have reported syntheses of large-area single-crystalline Bi_2O_3 nanowires by atmospheric pressure CVD (APCVD) approach using bismuth tris(diethyldithiocarbamate) $[Bi(S_2CNEt_2)_3]$ as a precursor in the presence of oxygen via vapor–liquid–solid (VLS) mechanism. Gold-coated silicon wafer was used as a substrate for this purpose and the depositions were carried out in a quartz tube at 550°C. Scanning electron microscope (SEM) and TEM analyses show that the Bi_2O_3 nanowires have a diameter in the range of 50–100 nm and a length of up to tens of microns. The details of these nanowires are shown in the Figure 7.4. X-ray diffraction (XRD), energy-dispersive spectroscopy (EDS), high-resolution TEM (HRTEM) and selected-area electron diffraction (SAED) demonstrate that the nanowires are composed of pure tetragonal phase of β-Bi_2O_3 single crystal. The figure shows a low-magnification view of the large-area Bi_2O_3 nanowires. It clearly reveals that on the substrate a layer of wire-like products, which are slightly bent, and have a large aspect ratio with a length up to several tens of microns. The magnified view of the Bi_2O_3 nanowires shows that the nanowires have circular cross sections with almost uniform diameters of the order of 50–100 nm throughout their entire lengths. The clear lattice fringes with a *d*-spacing of 0.55 nm are consistent with that of the (110) planes of tetragonal Bi_2O_3, further confirming that the Bi_2O_3 nanowires are single crystalline in nature. No other nanostructures are reported for these oxides.

Though this material is sensitive to some specific gases, as reported earlier, no specific data on these sensors are reported for gas-sensing applications using nanostructured bismuth oxides at present.

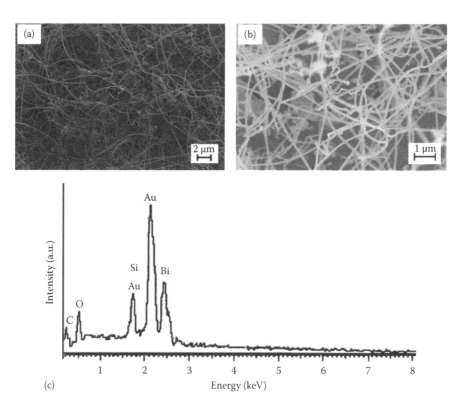

FIGURE 7.4
SEM images and EDS of the Bi_2O_3 nanowires. (a) Large area of the Bi_2O_3 nanowires, (b) magnified view of the Bi_2O_3 nanowires, and (c) EDS spectrum of the nanowires. (Reprinted from *Phys. E Low Dimens. Syst. Nanostruct.*, 39, Shen, X.-P., Wu, S.-K., Zhao, H., and Liu, Q., Synthesis of single-crystalline Bi_2O_3 nanowires by atmospheric pressure chemical vapor deposition approach, 133–136, Copyright 2007, with permission from Elsevier B.V.)

7.3 Cadmium Oxides

In its compounds, cadmium exhibits almost exclusively the +2 oxidation state and Cd^{2+} ion forms a large number of stable complexes. A few compounds of the +1 oxidation state are also reported for this material. The most important cadmium compound is cadmium oxide, CdO. For many other scientific applications, CdO compound serves as a base material.

Among transparent conductive oxide (TCO) materials, CdO shows promising prospect as a nanomaterial. This material exhibits *n*-type semiconducting properties and has a direct bandgap of 2.28 eV and an indirect bandgap of 0.55 eV. Among the TCO family, nanostructures based on CdO are particularly interesting because of their potential to serve as electrodes for nanoscale light-emitting diodes and lasers [59], an important issue for optoelectronic devices.

Another most important application of the micro-gas sensors is in the medical field, especially for human expiration and its evaluation. Careful analysis of the data provides important information on the state and functioning of different organs, decompensation of some pathologic states, or exacerbation of chronic diseases. Quantitative determination of oxygen and CO_2 in the expiration characterizes gas-interchange functions of the blood and lungs. Determination of acetone (CH_3–CO–CH_3) concentration in the blood of a sick person, particularly those suffering from diabetes, is an important medical application. Typical values of measured concentration range of this acetone vapor are in the range of 0.1–10 ppm [71]. However, the measurement problem is complicated because of the presence of H_2O and CO_2 impurities as they hinder the detection of acetone molecule as well as the decrease of O_2 concentration. Usage of sensors based on CdO provides a way for these applications, as these materials are sensitive for the gaseous species mentioned earlier. With proper signal analysis and estimations, cross-sensitivity factors, arising from other species, have to be minimized to get the reliable information from the human expiration.

These oxide films are generally deposited from water suspensions, by using finely ground powder, on alumina substrates [71]. Subsequently, the films are dried and thermally treated in dried air to use them as sensing elements. Several other techniques have been used to prepare polycrystalline CdO nanowires, both as single crystalline nanobelts, and microwhiskers; however, investigations on the electrical, optical, and chemical properties of high-quality, single-crystalline CdO nanostructures are still lacking to this material. Beside colloidal solutions or glasses, zeolite hosts can be excellently used as solid supports for the preparation of nanoparticles. Due to the well-defined pore systems in zeolites, dispersions with very narrow particle size distributions were obtained for oxides of cadmium [216]. Up to a distinct extent, the particle sizes are adjustable by the conditions of preparation, that is, choice of the zeolite type, degree of metal ion exchange, quality of dehydration prior to sulfidation, or temperature of calcination. It seems to be attractive to incorporate other metal-oxide nanoparticles, which in general exhibit quantum size effects and are expected to be much more stable.

Srivastava et al. [61] have reported a different approach to grow CdO nanostructured powder particles. They have used a simple thermal evaporation technique under atmospheric pressure at 1000°C. Deposition on quartz substrates has lead to tubular-, cylindrical-, horse-shoe-, and semi-spherical-shaped particles. These morphologies are shown in Figure 7.5. The inset shows the prominent tubular growth. It is further reported that these morphologies are sensitive to temperature during the growth, because the process is entirely catalytic free followed with solid–vapor mechanism. The different shapes with smooth surfaces are well defined at the time of cooling with the reaction of cadmium and oxygen is quite uniform. These novel microstructures have elucidated crystallographically preferred

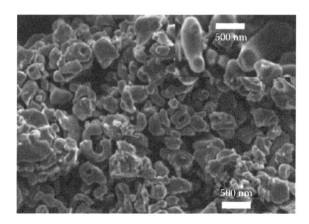

FIGURE 7.5
CdO sample showing morphologies of tubular, cylindrical, horse-shoe, and semi-spherical nanostructures. Inset shows prominent tubular growth. (Reprinted from *Mater. Lett.*, 62, Srivastava, A.K., Pandey, S., Sood, K.N., Halder, S.K., and Kishore, R., Novel growth morphologies of nano- and micro-structured cadmium oxide, 727–730, Copyright 2008, with permission from Elsevier B.V.)

growth along (002) crystal plane in a cubic lattice, instead of (111) plane generally observed in bulk CdO crystal formations.

Nanobelts of CdO with NaCl cubic structure were synthesized, by Wang [60], by evaporating CdO powders at about 1000°C. Besides CdO nanobelts, many single crystalline CdO sheets with sizes of the order of several to several tens of micrometers were also reported. These CdO sheets usually have different shapes of rectangles, triangles, and parallelograms. The lengths of the synthesized CdO nanobelts were reported to be less than 100 μm, and their widths typically in the range of 100–500 nm. Electron diffraction patterns showed that these nanobelts grow along the [100] direction and their surfaces are enclosed by ±(001) and ±(010) facets. Figure 7.6 shows TEM images of the as-synthesized CdO nanobelts.

Liu et al. [59] have synthesized and reported the electronic transport properties of CdO nanoneedles using CVD technique. Figure 7.7 shows SEM image of the needle-shaped with sharp tips about 40–100 nm in diameter and wide butts of several hundred nanometers to several micrometers at the other end, while the lengths ranged from 2 to ~20 μm. The crystal structure was found to be cubic with a lattice constant value of ~4.7 Å as determined by using XRD and SAED evaluation techniques. Devices based on individual nanoneedles were fabricated and studied, which revealed that the carrier concentration values of as high as 1.29×10^{20} cm^{-3} and thermal activation with energy of 13.3 meV was the dominant transport mechanism in these nano structures. The inset of the figure shows a magnified view of the butt end of one such nanoneedle, where a square cross section can be clearly seen. The formation was explained via the well-known VLS mechanism. Cadmium in the vapor phase first alloyed with

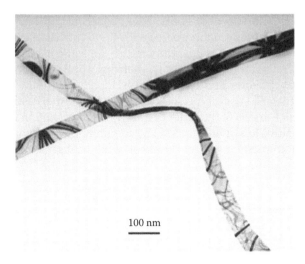

FIGURE 7.6
TEM images of the as-synthesized CdO nanobelts. (Reproduced with permission of Annual Reviews Inc. from Wang, Z.L., *Annu. Rev. Phys. Chem.*, 55, 159, 2004. With permission through CCC, Inc.)

FIGURE 7.7
SEM image of CdO nanoneedles grown using a CVD method. Inset: SEM image of a nanoneedle showing a square cross section. (Reprinted with permission from Liu, X., Li, C., Han, S., Han, J., and Zhou, C., Synthesis and electronic transport studies of CdO nanoneedles, *Appl. Phys. Lett.*, 82, 1950–1952. Copyright 2003, American Institute of Physics.)

the gold particles to form liquid droplets at high temperatures, and continued supply of cadmium vapor pushed the Au/Cd solution beyond the supersaturation. This led to nucleation of cadmium, which immediately reacted with oxygen and resulted in the growth of CdO nanowires along the longitudinal direction. These nanowire devices exhibited good sensitivity to both infrared light and also to diluted NO_2 gas, indicating their potential applications as infrared photodetectors and as toxic gas sensors.

Details of these CdO nanomaterials are explained in Table A.4. CdO deposition using CVD technique, nanoneedles with 40–100 nm diameter and 2–20 μm long structure, operation at room temperature are some of the important features for this CdO nanomaterial. For gas-sensing applications this material is sensitive to NO_2, acetone, and ethyl alcohol. Doped CdO, with 100 nm size, is an important material useful for detecting acetone. In this case, the response time is very quick and is of the order of 3 s only. In combination with SnO_2 this material is sensitive to alcohol vapors. The combination of acetone and alcohol is a major combination for the development of medical sensors.

7.4 Cerium Oxides

Cerium is different from the other trivalent rare earths in that it forms compounds in which it exhibits tetravalent behavior. Cerium is the only rare earth that shows +4 oxidation states in solutions. Tetravalent ceramic salts are powerful but stable oxidizing agents and are used in analytical chemistry to determine oxidizable substances. Most ceramic (IV) salts are orange to yellow in color. Cerium (III) behaves as a typical rare earth material but its compounds are usually white in color.

Cerium dioxide (CeO_2) or ceria-based oxygen sensors are currently being explored as possible candidates for oxygen sensing with applications especially in sensing automotive exhaust gases. Thin CeO_2 films exhibit very good adhesion, high thermal shock resistance and no apparent aging behavior under changing gas compositions at temperatures in the range of 700°C–1100°C. These CeO_2 sensors are suitable for closed-loop combustion control of stoichiometric or lean air-to-fuel mixtures.

Fast response of resistive type oxygen gas sensors, based on undoped nano-sized ceria powder was reported by Izu et al. [63]. This is based on porous thick film with the average nanoparticle size of the order of 100 nm. The material was synthesized by mist pyrolysis technique. Firing of this CeO_2 fine powder was carried out at 1100°C for 2 h in air prior to sensor fabrication. Sensors were fabricated by screen printing technique by using the powder paste onto the alumina substrates. Two methods were used to evaluate the response time of these sensors. The response time for sensors using these films with the particle size of 100 nm was shorter than that for corresponding sensor using films with the particle size of 200 nm to oxygen partial pressure dropped from a value of 10^5–10^3 Pa at the temperature of 615°C. It is further reported that the response times for the sensor with 100 nm particle were less than 1 and 0.25 s at 800°C and 900°C, respectively, from the results obtained by the method of periodic variation of oxygen partial pressure. It is expected that response time of these sensor based on CeO_2 thick film could be reduced by decreasing the particle size of the thick film. Subsequently, it was further reported by the same team [218] that the response times were unaffected by particle size of thick film but clearly depend on nanopowder firing temperature and strongly on the crystallite size.

Non-stoichiometric cerium dioxide (CeO_x, $1.5 < x < 2$) is an n-type semiconducting material with fluorite structure. Under normal conditions, typically encountered in auto exhaust, dissociation of CeO_2 as CeO_x and O_2 occurs easily and reversibly and CeO_2 accommodates a large number of oxygen defects, in the form of vacancies, while maintaining its fluorite structure. It is this reaction between the oxygen vacancies and the O_2 gas that leads to the changes in electrical conductivity of the sensing element.

Pure non-stoichiometric ceria, CeO_{2-x} is also an n-type semiconductor whose predominant ionic defects are oxygen vacancies in oxygen partial pressure of 10^5 Pa or lower. It is particularly suited as a sensor material because of its good structural stability and the associated high oxygen mobility. The material is generally prepared by chemical process in fine powder form. Figure 7.8a is the SEM micrograph of ceria powder with an average particle size of less than 50 nm. Figure 7.8b is the SEM micrograph of screen-printed thick film. In comparison, screen-printed film shows the appreciable grain growth, with an average size of around 100 nm, when compared to the as-prepared powder. The individual particles are clearly visible in the figure. n-type ceria has the relationship

$$\sigma = \sigma_o p(O_2)^{-r}, \quad \text{for } r > 0$$

where
 σ is the electrical conductivity
 σ_o is a constant value
 $p(O_2)$ is the oxygen partial pressure

FIGURE 7.8
(a) The SEM micrograph of the as-prepared ceria powder. (b) The SEM micrograph of the screen-printed thick-film sensor element. (Reprinted from *Sens. Actuators*, B89, Manorama, S.V., Izu, N., Shin, W., Matsubara, I., and Murayama, N., On the platinum sensitization of nanosized cerium dioxide oxygen sensors, 299–304, Copyright 2003, with permission from Elsevier Science B.V.)

Therefore, in oxygen partial pressure of the order of 10^5 Pa or lower, it is reported by Manorama et al. [3], that the output of the CeO_2:0.5% Pt sensor is dependent on oxygen partial pressure. The noble metal improves the response of the oxygen sensor significantly and definitely has a role in the sensitization mechanism for oxygen sensing.

Spinodal phase separation occurs when a binary system containing two phases with limited solubility is progressively concentrated. When both phases become immiscible, domains of one pure phase starts to form homogeneously within the system leading to the so-called spinodal phase separation in which domain size, morphology, and dispersion depend on the rate of extend of specie concentration and their diffusion in the present medium. If this phenomenon takes place in the presence of a condensable inorganic phase, texturation becomes possible. Original and homogeneous macro-textures shaped with coral-like, helical, or macroporous sieves morphologies have been obtained following a nanotectonic approach based on the template-directed assembly by poly-γ-benzyl-L-glutamate (PBLG) of organically functionalized CeO_2 crystalline nanoparticles. By adjusting a single parameter, such as the template to inorganic ratio, a versatile tuning between templating effect and phase separation yields hierarchical porous materials presenting both micro- and macroporosity with inorganic walls constituted of nanocrystalline CeO_2 particles [62]. Figure 7.9a shows the macroporous network obtained by using spinodal phase separation technique. Figure 7.9b is a tilted and enlarged view of the same network. This type of network exhibits a fine pore size around 1 μm and exposes large surface area for the interaction of oxygen gas with CeO_2 network.

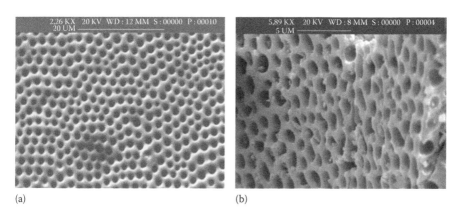

(a)   (b)

FIGURE 7.9
(a) CeO_2 macroporous network obtained by spinodal phase separation with PBLG. (b) Tilted and enlarged view of the network. (Reprinted from *Comp. Rend. Chim.*, 6, Sanchez, C., Soler-Illia, G.J.A.A., Ribot, F., and Grosso, D., Design of functional nano-structured materials through the use of controlled hybrid organic–inorganic interfaces, 1131–1151, Copyright 2003, with permission from Académie des sciences Elsevier SAS.)

The details of these CeO_2 nanomaterials are briefed in Table A.5. Presently, chemical syntheses are the only option for this material and are selective for oxygen species. The response time for these materials is very fast when compared with others. Nano-CeO_2 can improve the oxygen sensitivity of TiO_2 for its large special surface area and capacity of the storing and releasing oxygen while the atmosphere changes. Nano-CeO_2 material acted as an effective catalyst for the oxygen sensitivity of TiO_2 in stacked combination [219]. By making proper thin-film depositions on silicon substrate this material appears to be a good choice for the fabrication of integrated gas sensors suitable for automobile industries to study the air-to-fuel mixtures. For operation, the sensing elements require high temperature and electrical conductivity is measured in relation to oxygen partial pressures. Response time for these sensors is very fast and typically shows a value around 1 s. This is a major plus point favoring these cerium-based sensors for their application in the automobile industry.

7.5 Cobalt Oxides

Extensive oxidation occurs when larger pieces of cobalt are heated above 300°C in air. In fine powder form it ignites spontaneously at this temperature. In its compounds cobalt always exhibits a +2 or +3 oxidation state, although states of +4, +1, 0, and −1 are also known. Both the divalent and trivalent states form numerous coordination compounds, or complexes. The coordination number of these complexes is generally 6. Cobalt forms two well-defined binary compounds with oxygen: cobaltous, or cobalt (II) oxide, CoO, and tricobalt tetroxide, or cobalto-cobaltic oxide, Co_3O_4. The most stable phase of cobalt oxide is CoO, a magnetic insulator in which Co^{2+} has the $3d^7$ configuration. No theoretical calculations are available on surface electronic structure of this oxide material. Room temperature adsorption of O_2 gas indicates that the cleaved surface of CoO is virtually inert to this gas but defective surfaces interact very strongly [211]. Gases interact with lattice oxygen ions and reduce the surface by the creation of O-vacancy defects. Heavily reduced surfaces adsorb the interacting gas.

It is reported that the oxides of cobalt, in thin-film form, are sensitive to NH_3 vapor with minimum cross-sensitivity to hydrogen gas. It is very common that the materials that are sensitive to NH_3 gas are also sensitive to hydrogen gas. However, in the case of cobalt oxides it is an exception. Films of CoO_x are sensitive to NH_3 gas even at room temperature. This material is difficult to integrate with other devices, but is ideal for the fabrication of individual and independent sensing elements. Different nanostructured oxides of cobalt are synthesized by various techniques and at present this is one of the important nanomaterials under active study. Laser ablation is a popular technique to deposit these oxide nanoparticles. Other methods include spray pyrolysis and sol–gel techniques.

Tsuji et al. [65] have demonstrated the preparation of nanoparticles of CoO and Co_3O_4 by using laser-ablation technique, at an intensity of 30 mJ/pulse for 60 min, in H_2O and hexane (C_6H_{14}) solutions. It was found that drastic changes in the composition of these nanoparticles were observed upon the ablation. It was reported that composition of these obtained nanoparticles depended on the solvent in which laser ablation was carried out. This finding indicated that suitable selection of solvent is essentially an important criterion to control the composition of the nanoparticles prepared by using this ablation technique in liquids. TEM images of the nanoparticles obtained by laser ablation of Co, CoO, and Co_3O_4 are shown in Figure 7.10. The right images are expanded views of the left images for clarification of unclear smaller nanoparticles. This technique is attracting much attention as a new approach to prepare cobalt oxide nanoparticles.

Oh et al. [66] reported that nanocrystalline cobalt oxide (of phase Co_3O_4) powders, prepared via ultrasonic spray pyrolysis technique, with different crystallite sizes and morphologies at various calcination temperatures. This technique has unique advantage such as simplicity and low-cost alternative for the deposition of high-quality cobalt oxide thin films. The evolution of the crystal size (D_c) of this metal oxide, prepared at different temperatures, is shown in Figure 7.11. This graphical data shows that the growth in crystallite size has a strong relationship with the sintering temperature and directly results in different electrochemical performance. The crystallite size

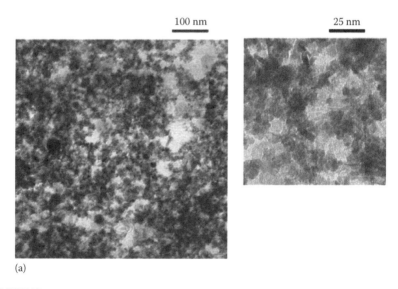

100 nm   25 nm

(a)

FIGURE 7.10
TEM images of nanoparticles obtained by laser ablation of various materials in water: (a) Co,

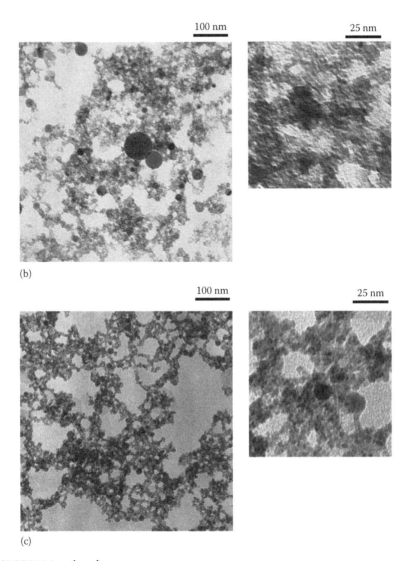

FIGURE 7.10 (continued)
(b) CoO, and (c) Co_3O_4. The right images are expanded views of the left images for clarification of unclear smaller nanoparticles. Laser ablations were carried out at 30 mJ/pulse for 60 min. (Reprinted from *Appl. Surf. Sci.*, 243, Tsuji, T., Hamagami, T., Kawamura, T., Yamaki, J., and Tsuji, M., Laser ablation of cobalt and cobalt oxides in liquids: Influence of solvent on composition of prepared nanoparticles, 214–219, Copyright 2005, with permission from Elsevier B.V.)

of Co_3O_4 increase linearly until the calcination temperature reaches 700°C. During this period, the variation in nanoparticle size varied from 20 to 62 nm over a wide temperature range of 300°C–700°C. After 700°C, the crystallite size slightly increases to 69 nm and shows a tendency toward saturation. Probably it may not improve further.

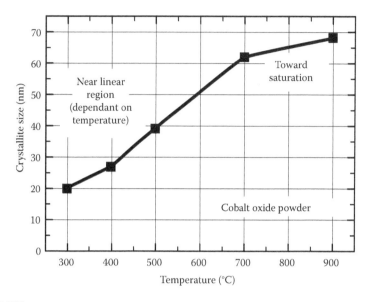

FIGURE 7.11
Evolution of crystallite size of Co_3O_4 powders sintered at different temperatures. (Reprinted from *J. Power Source.*, 173, Oh, S.W., Bang, H.J., Bae, Y.C., and Sun, Y.-K., Effect of calcination temperature on morphology, crystallinity and electrochemical properties of nano-crystalline metal oxides (Co_3O_4, CuO, and NiO) prepared via ultrasonic spray pyrolysis, 502–509, Copyright 2007, with permission from Elsevier B.V.)

A systematic sintering experiment was also carried out on these cobalt oxides. This is shown in Figure 7.12. These morphological studies were carried out, using SEM, at different sintering temperatures to record the physical changes. The samples of Co_3O_4 sintered at 300°C typically showed that the oxide has a spherical particle shape with a size distribution in the range of 0.5–2 μm. Between 300°C and 500°C, no significant morphological changes were observed. The particle size remained almost same during this temperature range. The morphology of samples, prepared at 700°C, started to disintegrate into smaller particles and reconvert back into spherical particles but with a smaller spherical shapes. It appears as though the larger spherical particles are disintegrating into smaller nanoparticles. This morphological change becomes clearer for Co_3O_4 prepared at 900°C. The spherical morphology of Co_3O_4 disappears totally and a new shape, with a larger particle size, was observed, as illustrated in Figure 7.12 at bottom side. This shape changes bring out the variation in surface area to interact with the gaseous species.

Gas-sensing properties of nanocrystalline Co_3O_4 in porous silica matrix are reported by Cantalini et al. [67] and were synthesized through sol–gel technique. Experimentally, it was proved the nanoporous structure of the silica network and the presence of crystalline metal-oxide nanoparticles dispersed uniformly inside the glass matrix. Electrical conductivity experiments on these films revealed a reversible change in total electric resistance,

FIGURE 7.12

SEM images of Co_3O_4 powders sintered at different calcinations temperatures. (Reprinted from *J. Power Source.*, 173, Oh, S.W., Bang, H.J., Bae, Y.C., and Sun, Y.-K., Sun, Effect of calcination temperature on morphology, crystallinity and electrochemical properties of nano-crystalline metal oxides (Co_3O_4, CuO, and NiO) prepared via ultrasonic spray pyrolysis, 502–509, Copyright 2007, with permission from Elsevier B.V.)

showing the *p*-type semiconducting behavior of these metal oxides, and in optical transmittance variation in the presence of reducing gases like CO and H_2. An operating temperature near 300°C was found to be the best range for the film, for CO as well as for H_2 detection. Detection limits of 10 ppm were demonstrated for CO and H_2 gases.

Details of the sensor properties are shown in Table A.6 for these cobalt oxide nanomaterials. Detection of carbon monoxide and hydrogen are the favorites of this cobalt oxide. Typical response time for these sensors vary between 1 and 7 min as represented in the table.

7.6 Copper Oxides

Copper forms compounds in the oxidation states +1 and +2 in its normal chemistry, although under special circumstances some compounds of trivalent are also reported. It was shown that this trivalent copper survives not more than few seconds in that particular chemical state. Copper (I) compounds are diamagnetic in nature and with few exceptions they are all colorless. Cuprous oxide (Cu_2O) is an important industrial compound and cupric oxide (CuO) has lot of commercial applications. CuO is also a promising material for fabricating solar cells due to the photoconductive and photochemical properties associated with this material.

Among the copper oxides, only cupric oxide (CuO) phase is reported as a gas-sensitive material and exhibits a number of interesting properties. Typically, CuO exhibits p-type semiconducting property, with a narrow bandgap of 1.2 eV, with low electrical resistance values. The electrical conductivity decreases, when it is exposed to the reducing gaseous species. In the ground state, CuO has a $3d^9$ electronic configuration and is an antiferromagnetic material with a bandgap of about 1.4 eV. The bandgap is identified as being of charge-transfer origin rather than arising from $d \rightarrow d$ transitions.

Nanoparticles, nanopowders, nanowires, and flower-like 3-D nanostructures are reported for these copper oxides by using different approaches to synthesize them. Reported techniques include: direct heating of copper metal in air ambient, ultrasonic spray pyrolysis, and through the typical chemical routes.

Copper oxide (CuO) nanowires are synthesized, by Xia et al. [68], by heating pure copper wire in air for 4 h at 500°C. SEM image of these fabricated nanowires are shown in Figure 7.13a. This is the simplest way to synthesize copper oxide nanowires. Two chemical reactions are involved in this process: first the oxidation of copper to form Cu_2O intermediate, and then a second oxidation step to generate CuO vapor from which uniform CuO nanowires are grown. Both electron diffraction and HRTEM studies confirmed that each of the CuO nanowire is a crystalline structure divided by a (111) twin plane in the middle along its longitudinal axis. Figure 7.13b shows the TEM images of CuO nanowires. Each CuO nanowire was a bicrystal as shown by its electron diffraction pattern and HRTEM characterization. Figure 7.13c is HRTEM image showing the twin boundary of a nanowire.

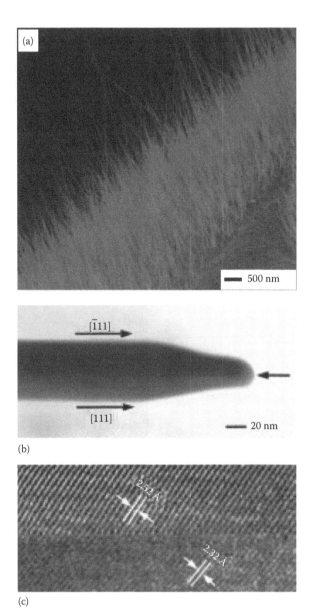

FIGURE 7.13

(a) SEM and (b) TEM images of CuO nanowires synthesized by heating copper wire, of 0.1 mm diameter, in air to a temperature of 500°C for 4 h. Each CuO nanowire was a bicrystal as shown by its electron diffraction pattern and HRTEM characterization. (c) HRTEM image showing the twin boundary of a nanowire. (From Xia, Y., Yang, P., Sun, Y., Wu, Y., Mayers, B., Gates, B., Yin, Y., Kim, F., and Yan, H.: One-dimensional nanostructures: Synthesis, characterization, and applications. *Adv. Mater.* 2003. 15. 353–389. Copyright Wiley-VCH Verlag GmbH & Co. KGaA. Reproduced with permission.)

Zhang et al. [69] reported synthesis of flower-like nanostructures of CuO 3-D flower nanostructures by utilizing template-free solution route on a copper surface. It is indicated that the amounts of chemical reagents play another important role in the formation of these structures. Reaction time plays a key role in the features as well. An SEM image of these flower-like features is shown in Figure 7.14. Figure 7.14a and b show the SEM images with different magnifications. When the reaction time is of short duration

FIGURE 7.14
SEM patterns with different magnifications of CuO 3-D nanostructures composed of microflowers prepared in different reaction times: (a) and (b) at 15 min, (c) and (d) at 25 min, and (e) and (f) at 40 min. (Reprinted with permission from Zhang, X., Guo, Y.-G., Liu, W.-M., and Hao, J.-C., CuO three-dimensional flowerlike nanostructures: Controlled synthesis and characterization, *J. Appl. Phys.*, 103, 114304. Copyright 2008, American Institute of Physics.)

(15 min), the product is mostly composed of grass-like structures and the microflowers. These microflowers are sparse and the diameter is relatively smaller (between 4.5 and 7.5 μm). As the reaction time is increased to 25 min, larger microflowers with diameter ranging from 6 to 9 μm are observed and their densities are also higher, as shown in Figure 7.14c and d. When the reaction time reaches to 40 min, the size of microflowers is in the range of 8–11 μm in diameter but their density is relatively low as seen in Figure 7.14e and f. As the reaction time is further increased, the microflowers grow large at the expense of the smaller quantities, producing bigger size and lower density. All these are driven by the tendency of the solid phase in the systems, which adjusts itself to achieve a minimum total surface free energy. It is further reported that these results are highly reproducible and the method is a simple and easily controlled route for producing various novel superstructures similar to the one reported here. These flowers show very porous structure and are having unique morphological features for possible applications in the field of chemical catalysis, sensor devices, electrochemistry, superconductivity, and super hydrophobic coating application areas.

Nanocrystalline CuO powders were also prepared via ultrasonic spray pyrolysis technique [66]. These oxide particles with different crystallite sizes and morphologies are obtained at various calcination temperatures. The evolution of the crystal size (D_c) of the oxide prepared at different temperatures is shown in Figure 7.15. From the data, it is apparent that

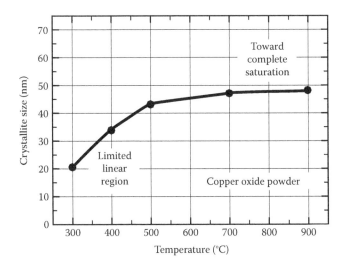

FIGURE 7.15
Evolution of crystallite size of CuO powders sintered at different temperatures. (Reprinted from *J. Power Source.*, 173, Oh, S.W., Bang, H.J., Bae, Y.C., and Sun, Y.-K., Effect of calcination temperature on morphology, crystallinity and electrochemical properties of nano-crystalline metal oxides (Co_3O_4, CuO, and NiO) prepared via ultrasonic spray pyrolysis, 502–509, Copyright 2007, with permission from Elsevier B.V.)

the growth has a strong relationship with the sintering temperature. The crystallite size of CuO increased linearly until the calcination temperature reached 500°C. It shows only limited linear region between the temperatures of 300°C and 500°C. At higher temperatures, greater than 500°C, CuO shows no significant change in its crystallite size, and saturates around 48 nm up to 900°C.

SEM micrographs of the CuO, obtained at different sintering temperatures, also show morphological changes with increasing temperature values. The samples prepared at sintering temperatures of 300°C and 400°C were observed to be spherical in shape and had near-identical morphological features with slight difference in their average diameter values. At 500°C, the spherical shape disappears completely and they exhibited totally different morphological features without any symmetry or shape. At this temperature of 500°C, the material gets fragmented into finer nanocrystalline particles. On further increasing the temperature to 700°C, recrystallization and agglomeration of the primary nanoparticles took place [66]. It is noteworthy to mention that the crystallite size of the CuO does increase much when compared to the samples of 500°C but the generic features are different at 900°C. At 700°C recrystallization is showing up and at 900°C the features are clearly visible. The SEM images of these CuO nanostructures are shown in Figure 7.16 as a function of sintering temperature. For clarity and comparison, all the SEM pictures are recorded at the same magnification value.

Copper oxide (CuO) catalyst and its dispersal in semiconducting oxides, particularly tin oxide (SnO$_2$), is gaining a lot of importance for trace-level detection of toxic hydrogen sulfide (H$_2$S) gas. Non-stoichiometric combination of CuO$_x$, along with SnO$_{2-x}$, which are p and n type semiconductors, respectively, have a strong electronic interaction due to which the CuO–SnO$_2$ thin-film surface consists of numerous p–n junctions thus causing very high electrical resistance of these films in air. These types of films are commonly synthesized by simultaneous vacuum evaporation technique. Initial values of these films were measured by Khanna et al. [33] and a resistance value was found to be of the order of 10^3–10^4 Ω only. On exposure to H$_2$S gas, CuO particles rapidly convert to CuS, which is metallic in nature and its formation destroys the p–n junctions existing on the surface thus causing a large decrease in the electrical resistance values. It is further reported that these thin-film sensors, fabricated on alumina disks, were capable of detecting few ppm levels of H$_2$S gas in atmospheric air and further reported that they exhibited reproducible results over a testing period of 3 years of operation. More studies are needed to confirm the reliability of these dispersed CuO sensors. This type of sensors is highly useful for the detection of H$_2$S gas at ppm levels.

Studies of CuO, with SnO$_2$ combination, were studied further by Chowdhuri et al. [220] and have reported about the CuO nanoparticles, based on SnO$_2$, for H$_2$S gas-sensing applications. Thin films were deposited on simple borosilicate glass substrates by using RF sputtering technique. These deposited

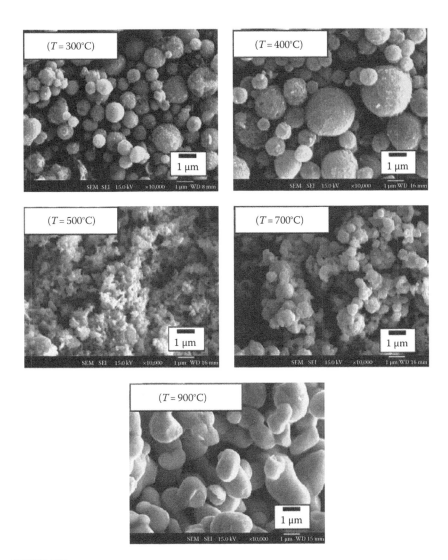

FIGURE 7.16

SEM images of CuO powders sintered at different calcinations temperatures. (Reprinted from *J. Power Source.*, 173, Oh, S.W., Bang, H.J., Bae, Y.C., and Sun, Y.-K., Effect of calcination temperature on morphology, crystallinity and electrochemical properties of nano-crystalline metal oxides (Co_3O_4, CuO, and NiO) prepared via ultrasonic spray pyrolysis, 502–509, Copyright 2007, with permission from Elsevier B.V.)

CuO nanoparticles exhibited a fast response speed with high sensitivity and showed a quick recovery time. The origin for the quick recovery is primarily due to the efficient dispersal of the CuO catalyst, and its presence in the form of nanoparticles proffers a greater surface area on the SnO_2 film surface. Figure 7.17 compares sensitivity and the response speed characteristics

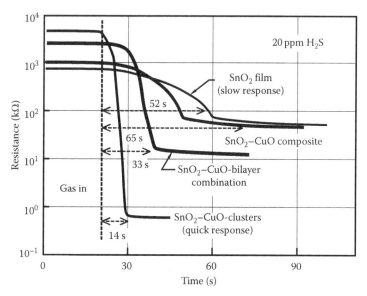

FIGURE 7.17

Sensitivity and response speed variation with dispersal of the CuO as a catalyst. (Reprinted with permission from Chowdhuri, A., Gupta, V., Sreenivas, K., Kumar, R., Mozumdar, S., and Patanjali, P.K., Response speed of SnO$_2$-based H$_2$S gas sensors with CuO nanoparticles, *Appl. Phys. Lett.*, 84, 1180–1182. Copyright 2004, American Institute of Physics.)

for trace level (20 ppm) H$_2$S gas detection for the first four sensors: (i) simple SnO$_2$ film layer (ii) mixed SnO$_2$–CuO composite layer, (iii) CuO/SnO$_2$ continuous bilayer, and (iv) evaporated CuO clusters on the SnO$_2$ structure. The SnO$_2$ film sensor with CuO clusters (0.6 mm diameter and 10 nm thick) exhibited high sensitivity, measured as 7.3×10^3 at a low operating temperature of 150°C, and a fast response speed of 14 s. An increasing CuO cluster thickness exhibited increasing values of R_a in the entire temperature range of 60°C–250°C. This is attributed to the increased penetration of the depletion region into the SnO$_2$ layer at the CuO/SnO$_2$ interface and effectively decreases the cross-sectional area of the conducting channel available for the charge carriers in the SnO$_2$ film.

Table A.7 shows the copper oxide nanomaterials for sensing application. It is observed that CuO is not directly used as a sensing material but in combination with SnO$_2$. In both the cases, it is in the form of thin films. This material is very specific for the toxic and inflammable H$_2$S gas and considered as a better choice for the detection. The detection limits are low and the response speed is very fast. The operating temperature is not very high when compared to other metal oxides. It is easy to create such temperatures, in the range of 130°C–200°C, using MEMS suspended platforms. Compatibility with standard silicon processing conditions and regeneration issues are not very clear.

7.7 Gallium Oxides

In most of its compounds, gallium shows an oxidation state of +3, however, in certain cases it shows +1 also. There is no authentic evidence for gallium (II) compounds. Metal gallium remains in the liquid phase over a wide temperature range between 0°C and 2000°C, with very low vapor pressure values up to about 1500°C, the longest useful liquid range of any element. Gallium slowly oxidizes in moist air until a protective film forms. It does not react with water at temperatures up to 100°C but reacts slowly with HCl and other mineral acids. Halogens attack it vigorously.

Gallium oxide nanostructures such as nanobelts [79], nanocolumns [79], nanopaintbrushes [76], nanoparticles [35], nanoribbons [78], nanosheets [79], nanotubes [76], and nanowires [75–79] were synthesized using different techniques. They include solvothermal treatment chemical process [35], physical evaporation from bulk gallium target [75], deposition using CVD technique [77], thermal annealing of compacted GaN compound [79], sublimation of gallium under argon gas environment in the presence of water vapor [78], and exposing molten gallium to appropriate composition of H_2 and O_2 in the gaseous phase reaction [76].

Presently semiconducting Ga_2O_3 films have been developed as a new material for gas-sensing applications because of their high-temperature stability. No other phase of oxides is reported for this specific purpose. This oxide phase exhibits an *n*-type semiconductor property, at elevated temperatures, and its semiconducting behavior is based mainly on the oxygen deficiency of the crystal lattice. These resulting oxygen vacancies stay ionized and form donors. β-Ga_2O_3 phase has a monoclinic crystal structure and has a wide bandgap of about 4.9 eV. Its remarkable thermal and chemical stability make it a suitable material for many technological applications, such as high-temperature oxygen sensor and magnetic tunnel junction, and also a UV-transparent conductive material.

This Ga_2O_3 material has been successfully employed in the form of sputtered polycrystalline thin films for gas detection. The material exhibits same charge carrier mobility in both the monocrystalline and in polycrystalline states [221], that is, electron mobility is independent of its grain boundaries [222]. The reproducibility of this critical quantity, and consequently its effect on the basic conductivity of the sensor, is considerably higher. It is also found that the sensors based on Ga_2O_3 films show high reproducibility and stability in the gas-sensitive electrical properties.

Not many nanostructures are reported in this important material. However, materials using solvothermal reaction between metal alkoxides and benzyl alcohol enables the synthesis of gallium oxide materials, and is reported by Niederberger et al. [35] by using nonaqueous and halide-free procedure. These as-synthesized nanoparticles showed agglomerated particles ranging from 50 to 80 nm and also exhibit good crystalline property,

FIGURE 7.18
Transmission electron micrograph of Ga₂O₃ nanoparticles together with the electron diffraction pattern. (Reprinted from *Prog. Solid State Chem.*, 33, Niederberger, M., Garnweitner, G., Pinna, N., and Neri, G., Non-aqueous routes to crystalline metal oxide nanoparticles: Formation mechanisms and applications, 59–70, Copyright 2005, with permission from Elsevier Ltd.)

as shown in Figure 7.18. A TEM image of γ-Ga$_2$O$_3$ phase is shown in Figure 7.18 together with the corresponding electron diffraction pattern. Here the particle size ranges from 2.5 to 3.5 nm.

Zhang et al. [75] reported the synthesis of Ga$_2$O$_3$ nanowires using physical evaporation technique using bulk gallium target under controlled atmospheric conditions. No separate oxygen source was provided for this process. The growth of these nanowires was about 60 nm in diameter but grew up to a whopping length of about 100 μm. They exhibited monocrystalline property, as analyzed by selected-area electron diffraction analysis, and are identified as monoclinic Ga$_2$O$_3$ structure with space group of $C2/m$. General morphology of these nanowires is shown in Figure 7.19. The finest nanowires are around 10 nm in diameter. Statistical counting of more than 300 nanowires has revealed an average diameter of about 60 nm of these nanowires as mentioned earlier. The lattice parameters of this crystalline phase are measured as $a = 1.223$ nm, $b = 0.304$ nm, $c = 0.580$ nm, and $\beta = 103.7°$, respectively. Figure 7.19a shows the low-magnification TEM image of gallium oxide nanowires prepared by physical evaporation; Figure 7.19b is a magnified TEM image revealing the general morphology of these nanowires, and Figure 7.19c is the histogram revealing the diameter distribution of these nanowires. Since no separate oxygen source was provided, the most likely source of oxygen, in this case, is likely from the silicon oxide of the quartz boat where metal was kept. The gallium is a metal, while silicon is a semiconducting material, so the former is more reductive in

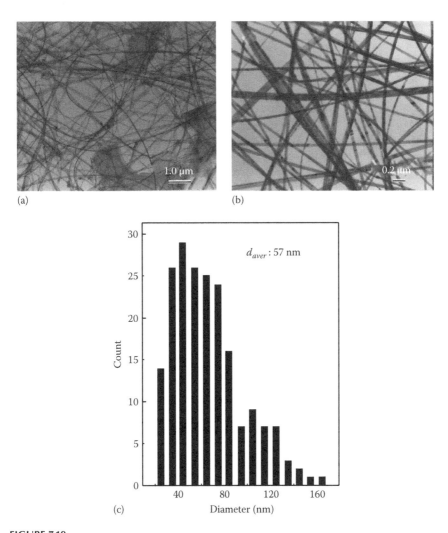

FIGURE 7.19
(a) Low magnification TEM image of gallium oxide nanowires prepared by physical evaporation. (b) Magnified TEM image revealing the general morphology of nanowires. (c) Histogram revealing the diameter distribution of the nanowires. (Reprinted from *Solid State Commun.*, 109, Zhang, H.Z., Kong, Y.C., Wang, Y.Z., Du, X., Bai, Z.G., Wang, J.J., Yu, D.P., Ding, Y., Hang, Q.L., and Feng, S.Q., Ga_2O_3 nanowires prepared by physical evaporation, 677–682, Copyright 1999, with permission from Elsevier Science Ltd.)

nature, and is easier to get oxygen from the silicon oxide on the surface of the quartz boat to form Ga_2O_3 nanowires. This type of long nanowires is useful for the development of micro-gas sensors where it is practical to place these nanowires between two metallic conductors and is easy to operate them in resistive mode. Such structures are also reported by Yu et al. [78] using single crystalline nanowires and nanoribbons.

Feng et al. [77] reported CVD route to synthesize β-Ga$_2$O$_3$ nanowires, which are highly sensitive to oxygen under ultraviolet radiation. They have demonstrated that these nanowires exhibited quick response to oxygen gas under 254 nm UV illumination and their electrical transport properties varied, depending upon the concentration of oxygen present in the environment, to which they are exposed. Their results are shown in Figure 7.20. The Figure 7.20a indicates that, without the illumination of UV radiation, the measured current is ~26 pA and it varies slightly under different oxygen pressures. In contrast, under the UV illumination, the current increased rapidly to a value that reflects the level of the oxygen pressure

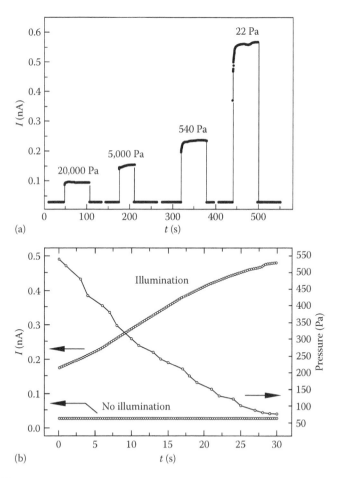

(a)

(b)

FIGURE 7.20
(a) Current–time curves under different oxygen pressures. (b) Variation of the current as the oxygen pressure changes. The illumination makes the nanowire sensitive to oxygen species. (Reprinted with permission from Feng, P., Xue, X.Y., Liu, Y.G., Wan, Q., and Wang, T.H., Achieving fast oxygen response in individual β-Ga$_2$O$_3$ nanowires by ultraviolet illumination, *Appl. Phys. Lett.*, 89, 112114. Copyright 2006, American Institute of Physics.)

in the chamber where the sensing element is kept. The measured current through the nanowire is ~0.56, 0.23, 0.15, and 0.095 nA as the oxygen pressure in the chamber is ~22, 540, 5,000, and 20,000 Pa, respectively. On UV illumination, the current is lower under higher oxygen pressure, as further demonstrated by measuring the current with continuously decreasing oxygen pressure. The comparisons in figures indicate very clearly that the nanowire is sensitive to the presence of oxygen only under the UV illumination. This is illustrated in Figure 7.20b. It is a clear indication that optically driven oxygen sensing is the origin of the fast response of this device. The ultraviolet illumination made the current across the nanowire increase rapidly to a value that reflected the oxygen pressure present in the chamber. The exact reaction mechanism is not known in this case. However, the room temperature operation and sensor quick response of this sensing element is a major point for oxygen detection in the environment. Deficiency of oxygen, in the working ambient, can be measured with this type of sensors and it appears that the sensors based on this oxide have better chances to lead in the race.

By thermal annealing of compacted gallium nitride (GaN) different nanostructures of β-Ga_2O_3 such as nanowires, nanobelts, nanosheets, and nanocolumns were synthesized by Jung et al. [79]. Decomposition of gallium from GaN and its subsequent reaction with oxygen in vapor form is proposed for the formation of Ga_2O_3. Some of the synthesized nanostructures are shown in Figure 7.21. Figure 7.21a shows the SEM images of β-Ga_2O_3 nanostructures and Figure 7.21b shows the plate-like hillocks grown on a c-plane sapphire, which was placed on a GaN pellet. The pellet was annealed at 900°C for 5 h. This study has demonstrated that the

(a)   (b)

FIGURE 7.21

SEM images of β-Ga_2O_3 (a) nanostructures and (b) plate-like hillocks grown on a c-plane sapphire, which was placed on a GaN pellet. The pellet was annealed at 900°C for 5 h. (Reprinted from *Phys. E Low Dimens. Syst. Nanostruct.*, 36, Jung, W.-S., Joo, H.U., and Min, B.-K., Growth of β-gallium oxide nanostructures by the thermal annealing of compacted gallium nitride powder, 226–230, Copyright 2007, with permission from Elsevier B.V.)

pellet method is a useful technique to synthesize β-Ga_2O_3 compound nanostructures without using any special catalyst. Besides nanowires, nanobelts, and nanosheets, different nano- and microcolumns were also obtained by the annealing of these GaN pellets. The Ga_2O_3 vapor diffuses into voids derived by compacting GaN powder and gets supersaturated in the voids, resulting in the growth of the Ga_2O_3 nanostructures through the vapor–solid condensation mechanism. The morphology of the nanostructures may be determined by the degree of supersaturation in these void formations. Ga_2O_3 plate-like hillocks and nanostructures were also grown on the surface of a c-plane sapphire substrate placed on the GaN pellet via the vapor–solid mechanism. The vapor is first condensed and then nucleated on the sapphire surface. Lateral and vertical growths from the nucleated seed result in hillocks and nanostructures, respectively. It is further pointed out that in these experiments Jung et al. [79] did not intentionally introduce oxygen in to the furnace, suggesting that the residual oxygen in the N_2 flow was the source of oxygen for this growth of Ga_2O_3 nanostructures.

Sharma and Sunkara [76] have synthesized highly crystalline Ga_2O_3 nanotubes, nanowires, and nanopaintbrushes using large gallium metal pools and a microwave plasma containing atomic oxygen. Direct use of gallium melts in plasma environments allowed bulk synthesis with high nucleation densities and allowed for template-free synthesis of nanostructures with unique geometrical features. Plasma excitation of the gas phase allowed for the synthesis of single-crystal quality nanostructures at much lower temperatures than that is commonly reported. In addition, the control of gas-phase chemistry allowed the manipulation of the nanostructure composition and morphology. Synthesized nanostructures of Ga_2O_3 are shown in the Figure 7.22. Figure 7.22a shows the one-dimensional (1-D) structures of a cluster of paintbrush-like structure of gallium oxide, Figure 7.22b is an example of individual nanopaintbrushes, and Figure 7.22c and d shows micrometer-scale nanotubes grown in the same experiment. These oxides exhibited 1-D structures with interesting morphologies. Micrometer-sized oxide tubes are seen with clear openings at the end points. The team has demonstrated the versatility of this process where highly crystalline structures were synthesized in bulk quantity and possible other growth routes for multiple nanostructures of gallium oxide.

Table A.2 explains the details of gallium oxide and different techniques adopted for the syntheses of nanostructures. Table A.8 shows the gallium oxide nanomaterials for gas-sensing application. Not many sensors are reported for these gallium oxide based on nanostructures. However, a two terminal nanowire device [78] operating at lower temperature, with a quick response of 2.5 s, appears to be a strong candidate for the detection of ethanol.

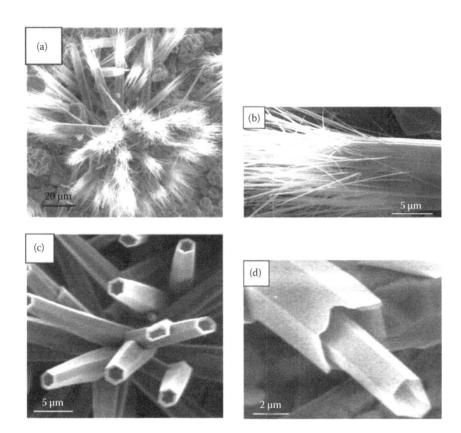

FIGURE 7.22
Gallium oxide 1-D structures with interesting morphologies. (a) Cluster of paintbrush-like 1-D structures of gallium oxide grown out of a gallium pool. (b) Individual gallium oxide nanopaintbrushes. (c, d) Micrometer-scale gallium oxide tubes grown in the same experiment. (Reprinted with permission from Sharma, S. and Sunkara, M.K., Direct synthesis of gallium oxide tubes, nanowires, and nanopaintbrushes, *J. Am. Chem. Soc.*, 124, 12288–12293. Copyright 2002, American Chemical Society.)

7.8 Indium Oxides

Though a few authentic indium(I) compounds have been reported, indium is commonly trivalent in its compounds. This metal is unaffected by air at ordinary temperatures, but at higher ranges it burns with a blue-violet flame to form a yellow colored oxide, In_2O_3. Indium metal is also unaffected by the boiling water but it readily dissolves in mineral acids. When heated in the presence of halogens or sulfur indium forms direct compounds.

Cubic indium oxide (In_2O_3) is a wide bandgap transparent semiconductor with relatively high electrical conductivity, in its non-stoichiometric form, and has been widely used in the microelectronic field as a winder heater, solar cell, and flat-panel display units. This is the only phase, among various oxides studied for gas-sensing applications and has attracted considerable research effort. The material has also attracted many researchers in the nanoforms such as ultrathin film, nano-size particle powder, and nanowires. This oxide is known to have a body centered cubic crystalline structure ($a = 10.12\,\text{Å}$) with a direct bandgap of 3.75 eV. It is well documented that the characteristics of In_2O_3-based sensors depend strongly on the conditions of their preparation, which determine the atomic structure formations, phase composition and indium electronic states in the sensing material.

Various techniques are adopted for the synthesis of indium oxide nano-structures, such as condensation from vapor phase [87], film deposition by high-vacuum thermal evaporation (HVTE) [28], heating indium metal chunks in a flowing argon atmosphere at normal and at very high-temperature ranges [84], high-intensity focused electron beam on indium metal nanowires [28], hot-wall CVD (HWCVD) [89], laser ablation [81,82,90], sol–gel (SG) spun coating [28], solvothermal treatment chemical process [35], vapor phase deposition technique [43,88], and wet chemical routes [80]. These details are listed in Table A.2 along with other oxide nanostructures.

Highly crystalline cubic In_2O_3 thin-film sensors have been successfully prepared by Cantalini et al. [28] by using HVTE and SG routes enabling the measurement of NO_2 gas sensing at sub-ppm levels. Both the films showed high sensitivity to NO_2 gaseous species at 250°C operating temperature. Figure 7.23a and b shows the SEM photomicrographs of these HVTE and SG films, respectively. The SG film shows a well-developed polycrystalline and highly porous microstructure composed of approximately spherical but randomly distributed nanoparticles with an average size of 19 nm. HVTE film, on the other

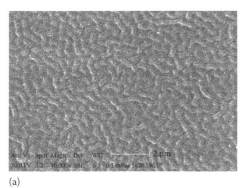

(a)   (b)

FIGURE 7.23
(a) SEM photomicrograph of indium oxide HVTE film after annealing 24 h at 500°C (marker 2 µm). (b) SEM photomicrograph of SG film after annealing at 500°C for 1 h (marker 0.2 µm). Note the difference in marker values. (Reprinted from *Sens. Actuators*, B65, Cantalini, C., Wlodarski, W., Sun, H.T., Atashbar, M.Z., Passacantando, M., and Santucci, S., NO_2 response of In_2O_3 thin film gas sensors prepared by sol–gel and vacuum thermal evaporation techniques, 101–104, Copyright 2000, with permission from Elsevier Science S.A.)

hand, is more dense and of coalescent grain with an "orange skin" appearance at lower magnification. It is further reported by Cantalini et al. [28] that SG prepared films show better NO_2 sensitive properties due to the formation of highly porous microstructure and reduced particle size with respect to HVTE prepared films. H_2O and C_2H_5OH interfering gases largely affect the electrical response in the presence of NO_2. SG films show improved selectivity to NO_2 gas in the presence of ethanol, with respect to the HVTE. Negligible H_2O cross has resulted in the range of 40%–80% relative humidity range. However, only 1000 ppm C_2H_5OH has resulted in a significant cross to the NO_2 response.

Shape-controlled nanowires and nanoparticles of In_2O_3 have been reported by Zhang et al. [84]. They have synthesized these nanostructures by heating indium metal chunks at high temperature in argon gas flow conditions, at the normal ambient pressure, in a simple CVD route. They have successfully grown different types of In_2O_3 nanowires, nanocubes, and octahedral-shaped particles of this metal oxide at nanometric dimensions. Evaluation of these structures by SEM showed that the obtained final nanomaterial is a mixture of nanowires along with some In_2O_3 particles. The diameters of these synthesized nanowires varied from 100 nm to several thousands of nm and the lengths were measured from several to hundreds of microns. In some locations of the CVD reactor tube, the deposited whiskers/nanowires were the main products. The whiskers showed a smooth surface morphology whereas the nanowires were spiral in nature, as shown in Figure 7.24a.

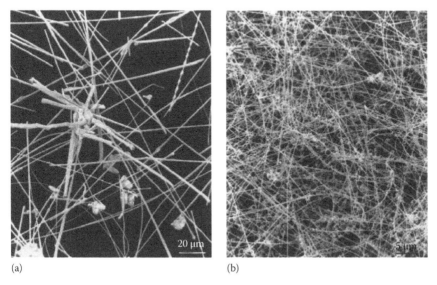

(a)   (b)

FIGURE 7.24
The morphologies of the indium (oxide) whiskers and nanowires. EDS of these nanostructures show that the main composition is indium having little oxygen. (Reprinted from *J. Cryst. Growth*, 264, Zhang, Y., Ago, H., Liu, J., Yumura, M., Uchida, K., Ohshima, S., Iijima, S., Zhu, J., and Zhang, X., The synthesis of In, In_2O_3 nanowires and In_2O_3 nanoparticles with shape-controlled, 363–368, Copyright 2004, with permission from Elsevier B.V.)

The pitch of these spiral structures is quite uniform. Some of these whiskers/ nanowires were found with a big droplet at one end of them while others are without such droplets. The reason for the droplets is not very clear. At locations the droplet is bigger than the average diameter of the nanowires resembling growing crystal growth from liquid phase. Details of these nanowires are shown in Figure 7.24b. EDS of these nanostructures showed that the main composition of these nanostructures is indium having little oxygen.

The main composition of these whiskers was found as indium metal used in the experiment with trace amounts of oxygen in it. Further the oxygen content of these nanowires was estimated to be far from the stoichiometric composition of In_2O_3. The compositions of the indium (the remaining is oxygen) in the edges and cores of these nanowires range in 48–61 at.% and 64–82 at.%, respectively, which indicates that the nanowires are in a metal-rich composition, especially in the core of the nanowires. Study of high-resolution electron microscope (HREM) indicated that an individual In_2O_3 nanowire is a single crystal configuration with bcc structure and it grows along the direction perpendicular to one of the {111} lattice facets. Only the atoms in the surface of the nanowire were reported to be somewhat disordered. The cubic particles, as shown in the Figure 7.25, on the other hand, are In_2O_3 with bcc structure, having 100–200 nm in the length of their edges [84]. A cube, which shows a hexagonal shape when the electron beam is parallel to its diagonal, has a selected-area electron diffraction (SAED) pattern of {111} facets of bcc structure. The octahedrons usually have size in the range of 60–300 nm, but mainly in 200–300 nm. SAED analysis further indicated that these octahedrons also have the bcc structure, similar to that of In_2O_3 cube. Generally, facets tend to form on the particles to increase the portion of the low-index planes and thus, for the particles smaller than 10–20 nm, the surface is a polyhedron. These cubes and octahedrons usually have {100} and {110} crystal facets and the particles are occasionally found to be larger than 100 nm in size. Figure 7.25a shows the TEM image of the indium oxide cubes. Figure 7.25b and c show an individual cube and its corresponding SAED pattern. Figure 7.25d shows the SEM microphotograph a high yield of octahedrons, and Figure 7.25e and f are big octahedron and its corresponding SAED pattern.

Cheng et al. [89] reported the octahedral nanocrystals, nanobelts, nanosheets, and nanowires of In_2O_3 using HWCVD technique, thus accessing different growth regimes. They also pointed out that the nanocrystals of these oxides show preferential growth directions, such as: nanocrystals in (111) orientation whereas nanobelts predominantly show (200) and nanowires (110) faces. These orientations are shown in the Figure 7.26 along with the FE-SEM images of grown crystals. The bottom portions of the figure illustrate the sketches with different crystalline faces. The figure clearly demonstrates the preferential growth direction of these indium oxides.

Extremely high sensitivity of ethanol vapor was reported by Feng et al. [80] using In_2O_3 nanocrystals, bounded by {100} facets, as the sensing materials.

These nanocrystals were synthesized via wet chemical route and the size of these nanocrystals typically varied from a few tens of nm to about 200 nm. The SEM picture, as shown in Figure 7.27, indicate the features of these nanocrystals. The composition of the film has different regular shapes. It was also found that these nanocrystals were mainly the formations of cubic structures. Figure 7.27a shows the SEM image of synthesized In_2O_3 nanocrystals, and Figure 7.27b shows the energy-dispersive x-ray spectroscopy (EDX) pattern of these nanocrystals. The sensors made from these In_2O_3 nanocrystals exhibited very high sensitivity parameters and very short response and recovery time values. Their study offers a promising material for the fabrication of ultrahigh

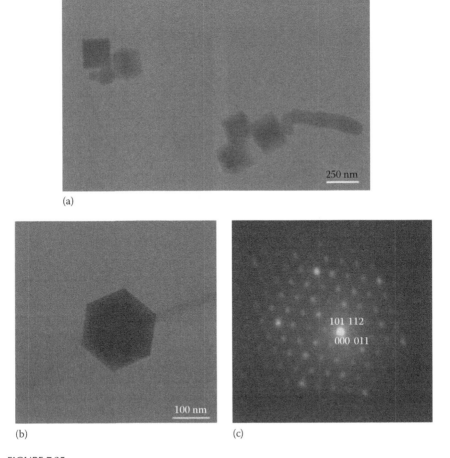

(a)

(b)   (c)

FIGURE 7.25
Indium oxide cubes and octahedrons. (a) TEM image of cubes. (b) and (c) an individual cube and its corresponding SAED.

(continued)

(d)

(e)   (f)

FIGURE 7.25 (continued)
(d) SEM microphotograph shows a high yield of octahedrons. (e) and (f) One (the big one) octahedron and its corresponding SAED. (Reprinted from *J. Cryst. Growth*, 264, Zhang, Y., Ago, H., Liu, J., Yumura, M., Uchida, K., Ohshima, S., Iijima, S., Zhu, J., and Zhang, X., The synthesis of In, In_2O_3 nanowires and In_2O_3 nanoparticles with shape-controlled, 363–368, Copyright 2004, with permission from Elsevier B.V.)
.

sensitivity ethanol sensors and also demonstrated a strategy for increasing the sensitivity of gas sensors particularly around 1 ppm concentration levels.

Nanowires and nanobelts were synthesized, by Comini [43], using vapor phase deposition, in a tubular furnace at higher temperatures, by suitably controlling the super saturation conditions inside the reactor. In this case, the deposition conditions are more critical due to the difficulties in producing anisotropic growth from symmetric structures. Different morphologies of In_2O_3 were obtained depending on the temperature of the substrates, varying from 1100°C down to 800°C, nanowires with lateral dimensions,

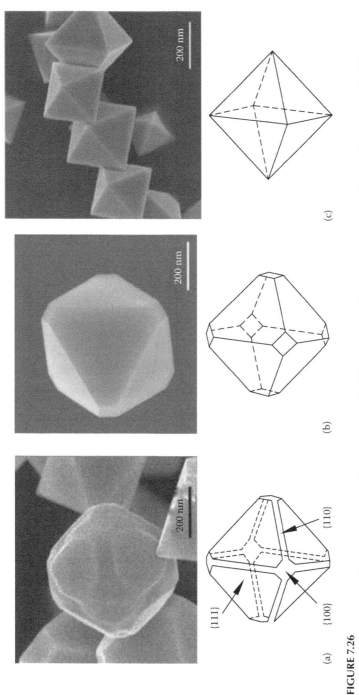

FIGURE 7.26

Different morphologies of octahedrons of In_2O_3 are shown in FE-SEM images, and the corresponding sketches are illustrated with different crystalline faces. (With kind permission of Springer Science+Business Media: *Appl. Phys. A: Mater. Sci. Process.*, Indium oxide nanostructures, A85, 2006, 233–240, Cheng, G., Stern, E., Guthrie, S., Reed, M.A., Klie, R., Hao, Y., Meng, G., and Zhang, L., 2006.)

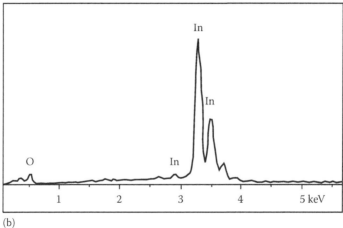

FIGURE 7.27

(a) Typical SEM image of synthesized In_2O_3 nanocrystals and (b) EDS pattern of these nanocrystals. (Reprinted with permission from Feng, P., Xue, X.Y., Liu, Y.G., and Wang, T.H., Highly sensitive ethanol sensors based on {100}-bounded In_2O_3 nanocrystals due to face contact, *Appl. Phys. Lett.*, 89, 243514. Copyright 2006, American Institute of Physics.)

of the order of several microns, were successfully grown. SEM images of these nanostructures are shown in Figure 7.28 for nanowires and 3-D crystalline structures. Studies using HRTEM indicated that these nanomaterials posses high degree of crystallinity. A cross section of these nanowires was found to be constant in all its length without any crystalline defect but with atomically sharp terminations. The figure shows the SEM images of different indium oxide structures obtained by a vapor phase deposition depending on substrates temperature: from 1100°C to 800°C wires with lateral dimensions of the order of microns, 3-D crystalline structures, and nanowires. This is an important achievement as the high degree of

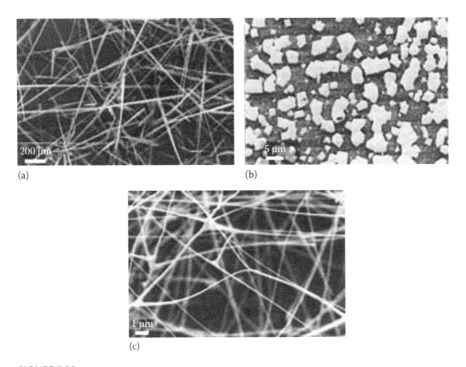

(a)   (b)

(c)

FIGURE 7.28
Secondary electron images of different indium oxide structures obtained by a vapor phase deposition depending on substrates temperature: from 1100°C to 800°C wires with lateral dimensions of the order of microns, 3-D crystalline structures, and nanowires. (Reprinted from *Anal. Chim. Acta*, 568, Comini, E., Metal oxide nano-crystals for gas sensing, 28–40, Copyright 2006, with permission from Elsevier B.V.)

crystallinity and the control in the surface termination is a key factor for producing and understanding stable and reliable gas-sensor devices. This type of perfect nanowires is preferred for the development of miniaturized gas sensors for field applications.

Niederberger et al. [35] developed novel reaction approaches using non-aqueous and halide-free procedures to synthesize indium oxide nanoparticles. Solvothermal reaction between metal alkoxides and benzyl alcohol enables the synthesis of a variety of binary metal oxides under mild conditions. In general, the as-synthesized nanoparticles are crystalline in nature and they exhibited uniform particle morphology with a narrow size distribution within one system. The particle shape of In_2O_3 is a cube-like structure with sides measuring about 20 nm. The details are shown in Figure 7.29. Evaluation of these particles has shown high degree of crystallinity as shown in the figure inset. It is further reported that the electrical measurement of the sensors, based on these materials, showed good response as a gas sensor. The electrical current dependence of gas flow showed that these In_2O_3 nanopowders exhibit higher sensitivity and better recovery time during gas-sensing measurement

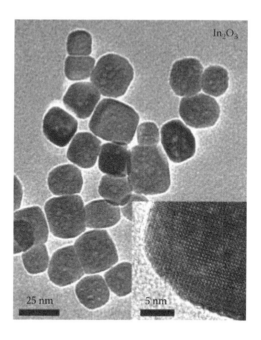

FIGURE 7.29
TEM micrograph of indium oxide cubic-like nanoparticles together with HRTEM image of single particle as inset. (Reprinted from *Prog. Solid State Chem.*, 33, Niederberger, M., Garnweitner, G., Pinna, N., and Neri, G., Non-aqueous routes to crystalline metal oxide nanoparticles: Formation mechanisms and applications, 59–70, Copyright 2005, with permission from Elsevier Ltd.)

in comparison with their bulky counterparts. Especially, these nanoparticles were highly sensitive toward NO_2 with a detection limit of 1 ppb at low temperature, a crucial detection range for this important polluting gas.

In_2O_3 is known to be an *n*-type semiconductor in its non-stoichiometric form mainly due to the presence of oxygen vacancies. This vacancy concentration acts like a doping in semiconductors. Its low-dimensional structures, such as In_2O_3 thin films and nanowires, have been extensively studied for the development of miniaturized gas-sensing devices. The electrical conductance modulation of these In_2O_3 is usually attributed to electron transfer between the sensing material and the gas molecules to be detected. This could be as oxidizing (or reducing) species that can serve as electron withdrawing (donating) groups and thus altering the carrier concentration of semiconducting In_2O_3 [81]. It is believed that photogenerated holes in In_2O_3 nanowires recombine with OH^- groups and oxygen ions adsorbed on the nanowire surface, and lead to the desorption of such species. Through the aforementioned process, electrons originally withdrawn from the nanowire by adsorbed oxygen or OH^- species were released back to the nanowire and the device reaches back to the original electron-rich state.

Li et al. [82] and Zhang et al. [81] studied the sensing behavior of NH_3 gas by using In_2O_3 nanowires. For this, they have used nanowires synthesized by using

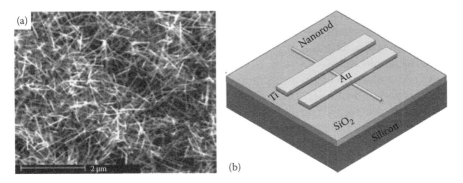

FIGURE 7.30
(a) SEM image of In_2O_3 nanowires grown by laser ablation on a Si/SiO_2 substrate coated with 20 Å of gold film. (From Li, C., Zhang, D., Han, S., Liu, X., Tang, T., and Zhou, C.: Diameter-controlled growth of single-crystalline In_2O_3 nanowires and their electronic properties. *Adv. Mater.* 2003. 15. 143–146. Copyright Wiley-VCH Verlag GmbH & Co. KGaA. Reproduced with permission.) (b) Picture shows the schematic diagram of In_2O_3 nanowire (rod) sensor on Si/SiO_2 substrate.

laser-ablation technique. These nanowires showed high uniform geometries and also exhibited single-crystalline property. Figure 7.30a shows the SEM image of these synthesized nanowires. A HRTEM image of the nanowire indicates that they have the single crystalline property. By using these In_2O_3 nanowires simple transistor behavior was studied. This is similar to the open-gate configuration with both ends behaving like source and drain of a typical field-effect transistor (FET) structure. In addition to the doping-dependent response to NH_3 gas, another interesting behavior exhibited by these In_2O_3 nanowire sensors. It is reported that the nanowire transistor sensors, as shown in Figure 7.30b, which is connected between two conducting Ti/Au contacts. When the whole structure is exposed to high concentration of NH_3 gas, their gate effect namely, the efficiency of gate bias to modify the channel conductance, substantially weakened by the adsorbed NH_3 species. This "gate-screening effect" was observed at high NH_3 gas concentrations. It is observed that chemisorbed NH_3 molecules screen the gate electric field by working as a charge trap.

This type of effect at room temperature and at a high concentration values (of the order of 1%) on SiO_2/Si substrates is a strong point for battery operated sensors for field and industrial applications. This type of configuration, using SiO_2/Si substrate platform, also helps for the development of integrated micro-gas sensors. Handling these nanowires is the crucial issue for such developments. Similar efforts are being explored for silicon nanowire transistors with circular gate-all-around structures [223–225].

In general, In_2O_3 nanostructures are used as an intrinsic material without any dopants due to its high electrical conductivity. However, it is reported [18] that the sensitivity and selectivity of these sensors, to oxidizing gases, can be improved further by doping with certain transition metal ions. In_2O_3 nanowires have been doped with gallium to tune further for better carrier concentration. It is reported that by tuning the concentration properly, electrical transport and gas-sensing properties were optimized. In addition, In_2O_3 nanowires have also been doped with tin, resulting in indium tin oxide (ITO) nanowires. ITO is a popular transparent electrical conductor in many scientific applications.

By using electron beam evaporation technique Tamaki et al. [4] fabricated In_2O_3-based thin-film sensors on SiO_2/Si substrate equipped with comb-type gold microelectrodes. The pure In_2O_3 film consisted of small grains that were stacked densely. The grain size was distributed in the range of 50–70 nm. On the other hand, the smaller grains in the range of 20–40 nm were observed for $In_2O_3(Fe_2O_3)$ film. The grain size of the oxide was decreased for the $In_2O_3(Fe_2O_3)$ thin film, indicating the enhancement of surface area of the film. The detection of dilute Cl_2 gas, of the order of 5 ppm, was realized by using the $In_2O_3(Fe_2O_3)$ thin films as sensing elements. Figure 7.31 shows the resistance and sensitivity of $In_2O_3(Fe_2O_3)$ thin-film sensor to 5 ppm Cl_2 gas as a function of sensor operating temperature. Temperature dependence of R_a is almost the same as that of pure In_2O_3 thin film, while R_g exhibited the maximum at about 400°C. Further the $In_2O_3(Fe_2O_3)$ sensor showed the highest sensitivity of ca. 70 at 400°C. Beyond this there is a sharp fall in the R_a value, clearly indicating the limited range of narrow temperature

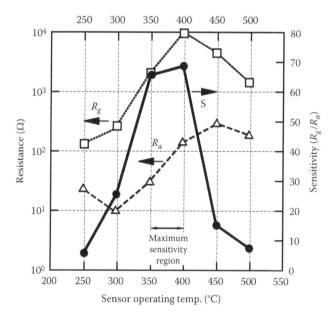

FIGURE 7.31
Resistance and sensitivity to 5 ppm Cl_2 gas of In_2O_3–Fe_2O_3 thin-film sensor as a function of operating temperature. (Reprinted from *Sens. Actuators*, B83, Tamaki, J., Naruo, C., Yamamoto, Y., and Matsuoka, M., Sensing properties to dilute chlorine gas of indium oxide based thin film sensors prepared by electron beam evaporation, 190–194, Copyright 2002, with permission from Elsevier Science B.V.)

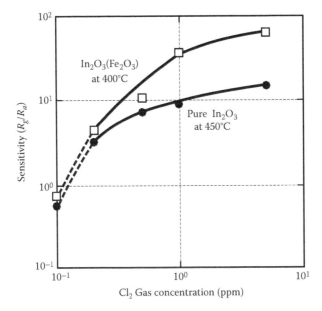

FIGURE 7.32

Sensitivities of In_2O_3 and In_2O_3–Fe_2O_3 thin-film sensors as a function of Cl_2 concentration; (●) In_2O_3 at 450°C, (□) In_2O_3–Fe_2O_3 at 400°C. (Reprinted from *Sens. Actuators*, B83, Tamaki, J., Naruo, C., Yamamoto, Y., and Matsuoka, M., Sensing properties to dilute chlorine gas of indium oxide based thin film sensors prepared by electron beam evaporation, 190–194, Copyright 2002, with permission from Elsevier Science B.V.)

region for its operation. Figure 7.32 compares the Cl_2 sensitivities of pure In_2O_3 and In_2O_3–Fe_2O_3 sensors as a function of Cl_2 concentration. The sensitivity monotonously increased with increasing Cl_2 concentration for both the sensors. The sensitivity was less than unity to 0.1 ppm Cl_2, suggesting that the direction of response changed at this concentration. The sensitivities of In_2O_3 and $In_2O_3(Fe_2O_3)$ sensors are 3.3 and 4.5–0.2 ppm Cl_2, respectively. It is proved by Tamaki et al. [4] that the sensitivity to dilute Cl_2 was much improved by the addition of Fe_2O_3 to In_2O_3. Though the operating temperature was lower in the case of In_2O_3–Fe_2O_3 the sensitivity is higher due to doping of indium oxide.

The details of the indium oxide nanomaterials are given in Table A.9. As it was pointed out in the table, most of the studies on these metal oxides are electrical in nature. Sol–gel process is popular, which explains that the chemical route is exploited thoroughly for this metal oxide. Thermal and e-beam evaporation techniques also gave 20–70 nm size grains. Operation of these devices was found to be in the range of room temperature and 350°C, except for the oxide doped with Fe_2O_3 that was operated up to 500°C. This oxide material, in nanodimensional structures, is sensitive to O_3, O_2, NO_2, C_2H_5OH, NH_3, in addition to Cl_2 gases ranging from ppb to ppm levels. In most of the cases the response time of these devices varied between 1 and 5 min. The ppb level detection, particularly to detect chlorine gas, is an important point worth noting for this indium oxide.

7.9 Iridium Oxides

Iridium is the most corrosion resistant metal known. It is not attacked by any acid, aqua regia, molten metals, or silicates at high temperatures. Iridium typically shows the oxidation states of +1, +3, and +4, though compounds of 0 to +6 are known except +2. A thick adherent oxide film, IrO_2, forms above about 600°C temperature. This oxide dissociates around 1100°C and above this temperature a volatile oxide, possibly IrO_3 forms. This material has a strong tendency to form coordination compounds.

Not much information is available on the iridium oxide materials and the nanostructures. This oxide crystallizes in the rutile structure and exhibits some interesting gas-sensing properties. Being the best conductor among the transition metal oxides this material shows the metallic conductivity at room temperature. Table A.10 shows the details of this material and also the FET structures for NO_2 sensing application [92]. The hybrid suspended gate FET, operating at 130°C is an important achievement for this material. The response time is too large for device operation.

7.10 Iron Oxides

The most important oxidation states of iron are +2 and +3. It also exhibits the states of +4 and +6. Iron(II) compounds are designated ferrous and contain Fe^{2+} ion and iron(III) compounds are called ferric and contain Fe^{3+} ion. Three oxygen compounds of iron are very common: iron(II) oxide or ferrous oxide, FeO; iron(III) oxide or ferric oxide, Fe_2O_3; and ferrosoferric oxide or ferroferric oxide, Fe_3O_4, which contain iron in both the oxidation states. Other stable form of iron oxide is Fe_xO that forms with non-stoichiometric compositions of iron and oxygen.

Chemistry and the bulk electronic properties of these iron oxides are quite complex, and they have received a great deal of attention both theoretically and experimentally. Fe_xO is a *p*-type semiconductor and very easily gets oxidized to form stable oxide. Fe_2O_3 is a magnetic insulator with the $3d^5$ electronic configuration. The mixed-valence inverse spinel Fe_3O_4 shows high electrical conductivity at room temperature and above, but at lower temperatures undergoes a much less conducting phase [211]. The valence state of iron plays very important role in exhibiting electrical conduction. α-Fe_2O_3 formations occur naturally as the mineral hematite where the crystals do not cleave. Their predominant growth faces are (0001) and (10$\bar{1}$1). Nearly stoichiometric α-Fe$_2$O$_3$ (0001) surfaces are quite inert for surface adsorption of gaseous species, probably due to the stability of the half-filled d-band configuration. Only weak interactions are possible with pure material. Defective surfaces or defect creation by incorporating other impurity species is the best way to use this material for gas-sensing applications.

α-Fe$_2$O$_3$ shows good structural and thermal stability and is resistant to photocorrosion. Besides its application as a gas-sensing material, this thermodynamically stable phase of iron(III) oxides is useful as a photoelectrochemical device. This is due to its resistance to photocorrosion, as referred earlier, high quantum efficiency as well as its low fabrication cost involved and availability in aplenty. As the most stable oxide phase under ambient condition, α-Fe$_2$O$_3$ is widely used for catalysts, nonlinear optics, and gas-sensor applications. This material shows an energy bandgap of 2.2 eV.

Table A.2 shows various techniques adapted to synthesize the nanostructures of Fe$_2$O$_3$. Nano-thin films were first deposited by the magnetron sputtering of base metal and its subsequent oxidation [71] is a popular technique. Different varieties of nanoshapes [72], flower-like nanostructures [73], and various nanoparticles [70] were synthesized, as reported in the open literature, by following different chemical routes. No other iron oxide phase has been reported for gas-sensing application other than the one discussed earlier.

α-Fe$_2$O$_3$ nanostructures were grown via simple oxidation of pure iron metal by exposing it to oxygen environment. Besides using thermal oxidation of pure iron, α-Fe$_2$O$_3$ nanobelts and nanotubes were produced from solution-based controlled wet chemical approaches. Growth of nanotube structures is also reported via a hydrothermal method [18]. The formation mechanism of tubular-structured α-Fe$_2$O$_3$ has been proposed as the coordination-assisted dissolution process. It is also reported that the Fe$_2$O$_3$ nanowire arrays, with an average diameter of about 120 nm and lengths up to 8 μm, were synthesized in AAO templates through electrodeposition and heat treatment of β-FeOOH.

Yan and Xue [73] adopted a new strategy to synthesize iron oxide nanostructure arrays, on an iron surface, by employing NaCl solution corrosion-based approach. This strategy has been successfully applied to grow nanofeatures and their approach is really excellent. They have selected iron metal foil as the base substrate for the growth of a well-orientated iron oxide arrays. This was used because of good lattice matching between the oxide and the metal to grow well-aligned nanostructures. Since, metal foil is an electrical conducting material and is very useful to utilize these aligned nanostructures for electronic and other device applications. With this simple technique of solution corrosion very interesting shapes were reported. Figure 7.33a shows a decorative flower shape Fe$_2$O$_3$ nanostructure is created by the formation of different Fe$_2$O$_3$ nanocrystals, where the continuous corrosion strategy is adopted in synthesizing them. Figure 7.33b is an enlarged picture of one such flower. The notable feature of this shape is the advantage of more surface area it offers, a desirable feature for gas-sensing applications.

(a)   (b)

FIGURE 7.33

(a) SEM image of flower-like nanostructures of Fe_2O_3. (b) Picture at a higher magnification of the structure. (Reprinted from *J. Cryst. Growth*, 310, Yan, C., and Xue, D., Solution growth of nano- to microscopic ZnO on Zn, 1836–1840, Copyright 2008, with permission from Elsevier B.V.)

Recently, Vayssieres [72] successfully demonstrated the growth of α-Fe_2O_3 nanostructures by using a novel approach on different substrates, which is based on thermo-dynamical growth control concept. This is a low temperature (95°C) aqueous thin-film growth technique, carried out at a large scale but, at a lower cost. Three-dimensional (3-D) arrays consisting of building blocks of controlled morphologies, sizes and orientations, engineered from molecular to nano-, meso-, or microscopic scales are generated onto various substrates such as ITO, single crystal silicon wafers, and sapphire. The details of SEM images on structural features are shown in Figure 7.34. These images of purpose-built crystalline α-Fe_2O_3 structures are grown directly onto various substrates of ITO, single crystal silicon and sapphire by controlled aqueous growth technique at a temperature of 95°C. The structures are different in shape and sizes. The surface condition of these substrates also influences their formation mechanisms. This figure illustrates the ability of the aqueous chemical growth technique to design the advanced purpose built nanomaterials.

For gas-sensing applications, Fe_2O_3 needs higher temperature and such temperature creation is difficult on silicon substrates and its compatibility with silicon processing conditions is not very promising. MEMS structure is an option for this material. In the intrinsic range, this material detects CO, CO_2, CH_4, O_2 and H_2 in the temperature range of 450°C–1075°C. The material, in its pure form, has very little to offer as a good gas sensor. When Fe_2O_3 is doped with Al_2O_3 (or La_2O_3) [228] the sensor operating temperature can be brought down to 420°C, where it is sensitive to different hydrocarbon gases, including LPG. With ZrO_2, it was reported by Tan et al. [206] that nonequilibrium nanostructured oxides give enormous oxygen dangling bonds, at the surfaces, and provide more active sites for the interaction of alcohol gas molecules for detection.

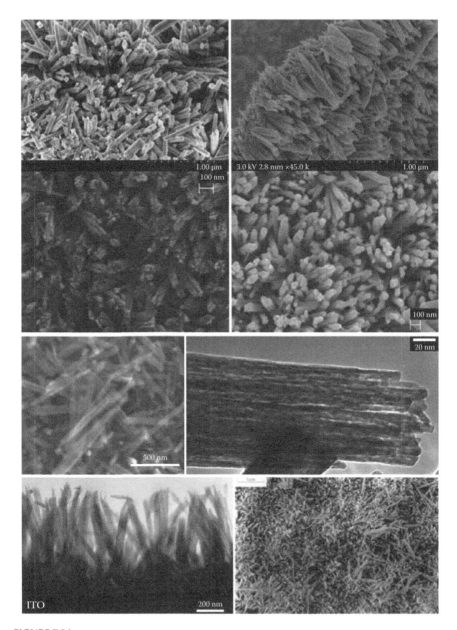

FIGURE 7.34
SEM images of purpose-built crystalline α-Fe₂O₃ grown directly onto various substrates of
ITO, single crystal silicon, and sapphire by controlled aqueous growth technique at a temperature of 95°C. (Reprinted from *Comp. Rend. Chim.*, 9, Vayssieres, L., Advanced semiconductor
nanostructures, 691–701, Copyright 2006, with permission from Elsevier SAS.)

Ryabtsev et al. [71] gave the details of these oxides, using Pt, Pd, or RuO_2 as active dopants, for this Fe_2O_3 to detect acetone at 300°C temperatures. Figure 7.35 shows the details of their studies on acetone, which is a key material for medical diagnostics particularly for diabetics. In this figure the sensitivity is showing a continuous raising trend with temperature. Operating the devices at lower temperature is an attractive issue for this sensing material. These materials are compatible to hybrid microcircuit fabrication technology and operation in the temperature range of 300°C–420°C, which is easily achievable in hybrid circuits with the help of external microheater structures kept below the ceramic substrates. Because of this advantage almost all the sensors based on Fe_2O_3 material are compatible to hybrid microcircuit technology using alumina as a base substrate for fabrication. However, the development of MEMS technology may offer better options in future.

α-Fe_2O_3 nanostructures have also been studied through configurable properties through doping procedures. To control their electrical properties,

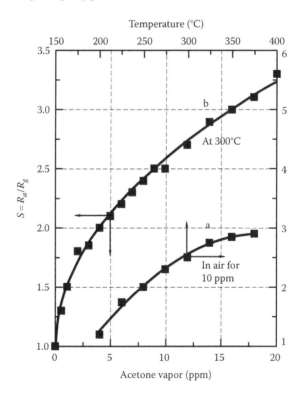

FIGURE 7.35
(a) Dependence of sensor response S with concentration of acetone vapor in air (10 ppm) for various temperatures. (b) Dependence of sensor response (at 300°C) on the concentration of acetone vapor in air. (Reprinted from *Sens. Actuators*, B59, Ryabtsev, S.V., Shaposhnick, A.V., Lukin, A.N., and Domashevskaya, E.P., Application of semiconductor gas sensors for medical diagnostics, 26–29, Copyright 1999, with permission from Elsevier Science S.A.)

α-Fe$_2$O$_3$ nanobelts were doped with elemental zinc [18]. Depending on the doping conditions, α-Fe$_2$O$_3$ nanobelts can be modified to either *p*-type or *n*-type with enhanced conductivity and electron mobility. Neri et al. [74] have shown that Zn and Au doping enhances the response to oxygen gas in the range of 10 ppm to 1%. Doping affects both the microstructure and the grain size of the sensing layer, forming a mixed oxide with iron.

Table A.11 shows the iron oxide nanostructures and the operating temperatures of the devices fabricated. It is clear that Fe$_2$O$_3$ is sensitive to acetone when doped with other dopant materials. Intrinsic material is not very popular for gas-sensing applications. Addition of other oxides, such as TiO$_2$, SnO$_2$, and ZrO$_2$, will enable this material to detect alcohol and other VOC. This is not a popular material for detecting other atmospheric pollutant gases. The operating temperature is also high for this oxide. The significant detection of acetone is a noteworthy point without any cross-sensitivity issues at ppm levels and probably this may draw larger attention in future.

7.11 Molybdenum Oxides

Molybdenum exhibits valences of +2 to +6 and is considered zero valent in the compound carbonyl, Mo(CO)$_6$. Chemically molybdenum dioxide and trioxide are the most common and are more stable. Molybdenum(VI) appears in the trioxide form from which most of its other compounds are prepared. Other oxides of molybdenum are metastable in nature and show a typical combination of MoO$_x$, where x has a value of 2.75, 2.8, 2.84, or 2.89. These molybdenum oxides show different colors, depending on the value of x, varying from wine-red, red-blue, blue-violet, violet, and blue-black. Surface instability is reported as one of the major issue with these oxides.

Molybdenum oxides are not very popular among the metal oxides for gas-sensing applications. Only MoO$_3$ phase is recorded for this application. This material has a bandgap of 3.2 eV and exhibits *n*-type electrical conductivity behavior. The resistivity of this material was measured to be of the order of 10^{10} Ω cm in its intrinsic form. The material exhibits a layered structure with orthorhombic symmetry. The crystal structure consists of double layer sheets. The most stable, basal plane is (010), on which only oxygen ions are exposed [211]. At room temperature, this exposed surface is quite inert. However, when the surface is exposed to electron or ion beams, even of relatively low energy, or to near-bandgap ultraviolet photons, the surface becomes reduced.

As a wide bandgap semiconducting material MoO$_3$ has received considerable attention over the last few years because of its many applications in various scientific fields. It is well known that MoO$_3$ acts as catalyst in chemical reactions involving hydrogen or oxygen molecules but only in the last few years MoO_3 thin films have attracted interest because of their application as the active element in conductance-type gas sensors. These sensors are found to be very sensitive to various gaseous species at relatively high operating temperature ranges. Syntheses of nanorods and nanotubes of MoO_3 are reported by using template directed hydrothermal process [93] and hot-wire CVD [94] techniques.

Taurino et al. [93] have synthesized single crystalline MoO_3 nanorods, as shown in Figure 7.36, by means of template directed hydrothermal process. Length of these nanorods ranged from 0.35 to 14 μm and the diameters from 20 to 280 nm. Study of the *I–V* characteristics, of these nanorods, showed that, at higher temperatures, current is thermally activated and successively it decreased with time. It is projected that this type of nanostructures may gain importance as a gas-sensing material, in near future, in view of their shapes. Suspended open structures, as shown in Figure 7.30b, are possible with this type of nanorods for the development of integrated micro-gas sensors.

Dillon et al. [94] have demonstrated HWCVD technique to synthesize this oxide. The HWCVD-generated material was a fine powder consisting of predominantly with nanorods, nanotubes with a few larger crystalline particles, and a small amount of amorphous structures. Both bulk single-wall nanotubes (SWNT) and multi-wall nanotubes (MWNT) were also present in the synthesized material. These nanotubes were found to be crystalline

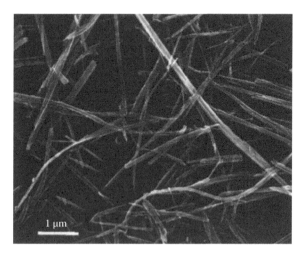

FIGURE 7.36
SEM image of MoO_3 nanofibers. Length of these fibers ranges from 0.35 to 14 μm and the diameters vary from 20 to 280 nm. (Reprinted with permission from Taurino, A.M., Forleo, A., Francioso, L., Siciliano, P., Stalder, M., and Nesper, R., Synthesis, electrical characterization, and gas sensing properties of molybdenum oxide nanorods, *Appl. Phys. Lett.*, 88, 152111. Copyright 2006, American Institute of Physics.)

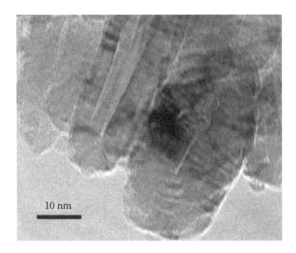

FIGURE 7.37
TEM images of MoO$_3$ nanotubes synthesized by HWCVD technique. (Reprinted from *Thin Solid Films*, 501, Dillon, A.C., Mahan, A.H., Deshpande, R., Alleman, J.L., Blackburn, J.L., Parillia, P.A., Heben, M.J., Engtrakul, C., Gilbert, K.E.H., Jones, K.M., To, R., Lee, S.-H., and Lehman, J.H., Hot-wire chemical vapor synthesis for a variety of nano-materials with novel applications, 216–220, Copyright 2006, with permission from Elsevier B.V.)

rods with ~10–20 nm diameters and ~40–60 nm in length. Figure 7.37 shows a TEM image of HWCVD-deposited MoO$_3$ nanorods and tubes. The crystallite structure was confirmed with XRD and showed the presence of MoO$_3$ phase. However, the density of nanotubes or nanorods was not very high in this particular material formation using HWCVD technique.

Molybdenum doping as MoO$_3$ with other oxide material significantly enhances the sensor response to NO$_2$ gas molecules [17]. The mechanism was attributed to the chemical reaction with O$_3$ and NO$_2$ dissociation to reactive species. It is expected that Mo-cations promote the O$_3$/NO$_2$ dissociation to reactive species. This was significantly noted in case of MoO$_3$–In$_2$O$_3$ mixers and In$_2$O$_3$ films show the best results in terms of NO$_2$ response with their average grain size between 5 and 30 nm. The reason of the high response to oxidizing gases (O$_3$, NO$_2$) for nanocrystalline metal-oxide sensors is, though not well understood, it is believed that large surface-to-volume ratio, or large number of unsatisfied bonds in nano-materials is responsible for sensor response enhancement. Adsorption of electron acceptor gaseous species leads to band bending and the formation of a surface depletion layer due to capture of the free charge carriers at surface. For nanocrystalline semiconductors, nearly all-available conduction electrons are trapped in the surface states, which enhance the sensor response. Comparing In$_2$O$_3$ and MoO$_3$–In$_2$O$_3$ sensitive layers prepared by the hydrolytic sol–gel method showed that the best NO$_2$ and O$_3$ response is observed for nanocrystalline thin films. Comparison of the experimentally observed exponents with modeling showed that O_2^-/O^- should be the predominant species on the surface due to the ozone and NO_2 interaction with In_2O_3 at 150°C–420°C.

Two problems are identified for MoO_3 as a gas-sensing material. First, the material has a low evaporating temperature, permitting only low operating temperatures. Second, the material has a very high resistivity, making it a difficult material to realize as a gas sensor and to integrate with electronics [230]. Although, these two disadvantages have been identified, MoO_3 possesses good gas response since it has been used in the field of catalysis for oxidation reactions of hydrocarbons.

Table A.12 briefs the molybdenum oxide based nanomaterials. In addition to ozone, these oxides are also sensitive nitrogen oxides and ammonia gaseous species. Sol–gel technique has an advantage, over other methods, and is preferred for the preparation. Electrical conductivity measurements are carried out through interdigitated structure. Response of the sensors based on this oxide is slow. No separate sensing behavior, with other metal oxides, is reported on this material except in combination with In_2O_3. It is too early to predict different possible nanostructures, for sensor applications, of this oxide.

7.12 Nickel Oxides

Nickel is usually dipositive in its compounds, but it can also exist in the oxidation states of 0, +1, +3, and +4. Next to the +2 oxidation state, the +4 state is most common for this metal. Nickel(IV) oxide, NiO_2, is probably a definite compound but stoichiometrically it was never perfect. Besides simple compounds nickel forms a variety of coordination compounds or complexes.

Nickel oxides are also not very popular as gas-sensing materials. Only NiO phase is reported for sensing applications. This phase exhibits rocksalt structure and shows highest quality cleavage along the (100) plane where coulomb interactions are possible. The cleaved NiO surface is inert to O_2 gaseous species, however, they interacts strongly with the defect surface, depopulating the Ni $3d$ defect structure at the upper edge of the valence band [211]. Hydrogen will not bond to perfect (100) surfaces with either nickel or oxygen vacancies necessary for adsorption.

Only three techniques are reported for the synthesis of nano-nickel oxides. They are nanopowders by using ultrasonic spray pyrolysis technique [66], nanocrystals in porous SiO_2 matrix by sol–gel technique [67], and by using molecular beam deposition [95]. Studies on O_3-enhanced molecular beam deposition of NiO_x, on silicon substrates, and the variation of film stoichiometry to the gas-sensing properties indicate that the sensitivity of the film can be adjusted by controlling both the film stoichiometry and morphology of the film [95]. The details are listed in Table A.2 along with other metal oxides.

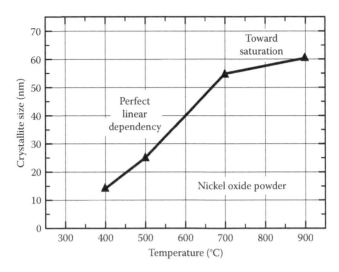

FIGURE 7.38
Evolution of crystallite size of NiO powders sintered at different temperatures. (Reprinted from *J. Power Sources*, 173, Oh, S.W., Bang, H.J., Bae, Y.C., and Sun, Y.-K., Effect of calcination temperature on morphology, crystallinity and electrochemical properties of nano-crystalline metal oxides (Co_3O_4, CuO, and NiO) prepared via ultrasonic spray pyrolysis, 502–509, Copyright 2007, with permission from Elsevier B.V.)

Nanocrystalline NiO powders were prepared, by Oh et al. [66], via ultrasonic spray pyrolysis technique with different crystallite sizes and morphologies. They were obtained at various calcination temperatures. The evolution of the crystal size (D_c) of the oxide prepared at different calcination temperatures is shown in Figure 7.38. Their data shows that the growth in crystallite size of nickel oxide powder has a strong relationship with the sintering temperature. The size of NiO increased linearly until the calcination temperature reached to a value of 700°C. In this range, the crystallite size increased from 14 nm (at 400°C) to 54 nm (at 700°C). It showed a perfect linear dependency in this range. Beyond 700°C, the crystallite size slightly increases to 60 nm, from 54 nm, and at 900°C the trend is more toward saturation in its size.

SEM study clearly shows the trend pictorially. Figure 7.39 shows the NiO powders sintered at different calcination temperatures, such as 400°C, 500°C, 700°C, and 900°C. As the sintering temperature is increased from 400°C to 900°C, the particle and agglomerate sizes remain unaltered. Between 400°C and 700°C, most of the particles exhibited a smooth spherical shape. On the other hand, NiO prepared at 900°C consists of particles with relatively coarse surfaces [66]. These particles are composed of submicron-sized, approximately 200 nm, primary particles that form secondary agglomerates, as shown in the figure. Though NiO phase is sensitive to different gases at high temperatures, nanostructured material sensing properties are not available for these nanopowders.

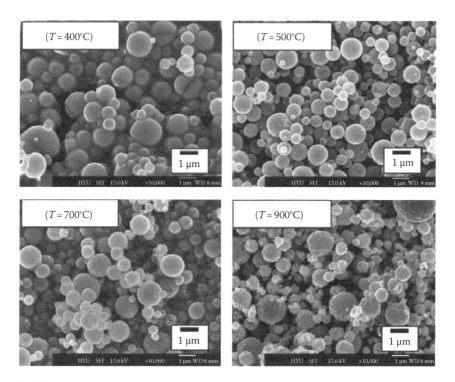

FIGURE 7.39
SEM images of NiO powders sintered at different calcination temperatures. (Reprinted from *J. Power Sources*, 173, Oh, S.W., Bang, H.J., Bae, Y.C., and Sun, Y.-K., Effect of calcination temperature on morphology, crystallinity and electrochemical properties of nano-crystalline metal oxides (Co_3O_4, CuO, and NiO) prepared via ultrasonic spray pyrolysis, 502–509, Copyright 2007, with permission from Elsevier B.V.)

Gas-sensing properties of nanocrystalline NiO in porous silica matrix are reported by Cantalini et al. [67] that were synthesized through sol–gel technique. Experimentally it was proved that the nanoporous structure of the silica network and the presence of crystalline metal-oxide nanoparticles dispersed uniformly inside the glass matrix. These films revealed a reversible change in their electric resistance, showing *p*-type semiconducting behavior. These metal oxides in the presence of reducing gases like CO and H_2 the NiO doped films showed a slightly higher sensitivity than the Co_3O_4 film in detecting CO gas, and a much stronger affinity toward H_2 detection than the Co_3O_4 doped film. An operating temperature near 300°C was found to be the best range for both the films, for CO as well as for H_2 detection. The NiO doped film also showed a better selectivity toward H_2 detection when CO was the interfering gas. Optical transmittance changes in the vis-NIR region occur when the films are exposed to CO, with the transmittance increasing as CO concentration increases. The detection limit when using optical transmittance as the sensor signal is better than 10 ppm CO.

Table A.13 shows the details of these nickel oxide nanomaterials. Compatibility of this oxide material to silicon dioxide, silicon nitride, and glass substrates is excellent and is a major technical point for the development of micro-gas sensors based on these nickel oxides. The response times of these devices, based on NiO, are rather slow.

7.13 Niobium Oxides

Compounds of niobium are of relatively minor importance. Its compounds found in nature are pentavalent, but compounds of lower valences have been reported. Its valence electron configuration is $4d^45s^1$, which accounts for its maximum oxidation state of niobium(V). Oxidation states of +2, +3, and +4 are also known. Although niobium has excellent corrosion resistance it is susceptible to oxidation above the temperatures of 400°C.

Niobium oxide (Nb_2O_5) is gaining popularity for gas detection applications and also very promising material for the development of integrated gas sensors. Nb_2O_5 belongs to the group of transition metal oxides, in which all the d electrons of the transition metal atoms are donated to the oxygen atoms. With a small excess of metal atoms compared to stoichiometry, Nb_2O_5 is an n-type semiconducting oxide [231] because of anionic vacancies. The detection principle of this material is based on the reversible modulation of the electrical conductance in the presence of oxidizing or reducing gases. This material has been especially tested as an oxygen sensor material where its conductivity decreases when oxygen partial pressure is increased.

Niederberger et al. [35] reported nonaqueous routes to synthesize crystalline Nb_2O_5 nanoparticles. The solvothermal reaction between metal alkoxides and benzyl alcohol enables the synthesis of a variety of niobium metal oxides under mild conditions. In general, the as-synthesized nanoparticles are crystalline in nature and exhibit uniform particle morphology with a narrow size distribution within one system. One such particle morphology of Nb_2O_5 is shown in Figure 7.40, where square-like platelets are formed. The discrete, non-agglomerated particles have sides ranging from 50 to 80 nm and are also of high crystallinity as shown in figure inset.

Table A.14 shows the Pt, Pd, RuO_2 doped Nb_2O_5 nanocrystallites, which are sensitive to acetone vapor, detectable at 300°C [71], which is an important observation. This material shows a response time of 3 s, for the detection of acetone in the range of 0.1–20 ppm, a rare feature among the metal-oxide nanostructures. This is the typical range for clinical observation.

FIGURE 7.40
TEM micrograph of Nb$_2$O$_5$ nanoparticles together with HRTEM image of single particle as inset. (Reprinted from *Prog. Solid State Chem.*, 33, Niederberger, M., Garnweitner, G., Pinna, N., and Neri, G., Non-aqueous routes to crystalline metal oxide nanoparticles: Formation mechanisms and applications, 59–70, Copyright 2005, with permission from Elsevier Ltd.)

7.14 Tellurium Oxides

The oxides of tellurium are tellurium monoxide, TeO; tellurium dioxide, TeO$_2$; and tellurium trioxide, TeO$_3$. The TeO is reported as a black and amorphous material. It is stable in dry air, in the cold, but gets oxidized in moist air to convert to TeO$_2$. On being heated in vacuum, TeO apparently disproportionates into the dioxide and elemental tellurium. On the other hand, TeO$_2$ is the most stable oxide. It is formed when tellurium is burned in air or in oxygen ambients.

TeO$_2$ is a versatile wide bandgap semiconducting material. Single crystal TeO$_2$ has wide applications in active optical devices such as deflectors, modulators, and tunable filters because of its remarkable acousto-optical and electro-optical properties. In thin-film form, TeO$_2$ has been studied as an optical storage material. TeO$_2$-based glasses find applications in laser devices and also in nonlinear optics. In the chemical front, TeO$_2$ is an important component of propane oxidation catalysts.

Liu et al. [124] demonstrated, for the first time, that TeO$_2$ single crystal nanowires are sensitive to the gases at room temperature. These nanowires were synthesized by thermal evaporation of tellurium metal in air, at 400°C, without using any catalyst. These nanowires have diameters ranging from 30 to 200 nm and several tens of micrometers long. Figure 7.41a shows the SEM image of the synthesized nanowires grown on silicon substrates. Figure 7.41b shows the XRD pattern of TeO$_2$ nanowires grown on silicon. This nanowire shows a well-crystallized tetragonal structure. Gas sensors were fabricated using these TeO$_2$ nanowires on oxidized silicon substrates showed good sensing performance to NO$_2$, NH$_3$, and H$_2$S gases at room temperature. These sensors worked in the detection range between 10 and 100 ppm. Room

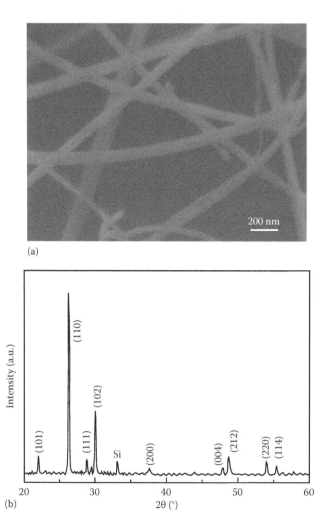

FIGURE 7.41
(a) Typical SEM image of TeO_2 nanowires grown on silicon substrate and (b) XRD pattern of TeO_2 nanowires grown on silicon substrate. The nanowires show a tetragonal structure. (Reprinted with permission from Liu, Z., Yamazaki, T., Shen, Y., Kikuta, T., Nakatani, N., and Kawabata, T., Room temperature gas sensing of *p*-type TeO_2 nanowires, *Appl. Phys. Lett.*, 90, 173119. Copyright 2007, American Institute of Physics.)

temperature detection of these pollutant gases and usage of silicon substrate for these TeO_2 nanowires may open new avenues for this material as a sensor material. Compatibility to silicon substrate is another advantage to use it for the development of integrated gas sensors.

Table A.15 shows the details of these TeO_2 nanowires about the detection limits and the response times for the pollutant gases. Sensors based on this nanomaterial are relatively slow to detect the gases mentioned.

7.15 Tin Oxides

Tin forms two series of compounds stannous or tin(II) compounds and stannic or tin(IV) compounds. Both are stable oxide formations. Stannous oxide, SnO, is a blue-black crystalline product. This compound is thermally stable in air up to 385°C, and converts to other form, that is, stannic oxide, SnO_2, a white-colored compound. In the chemical industries, SnO is employed in making tin salts for reagents and SnO_2 is popular in the petroleum industry as a heterogeneous oxidation catalyst.

Among the various semiconducting metal oxides, SnO_2 has been the most popular gas-sensing material, so far investigated, and used in practice. Great deal of research effort has been exerted to improve the gas-sensing properties of these SnO_2-based sensors and even today many scientific groups are working on this material. SnO_2 crystallizes in the tetragonal crystal system (space group $P4_2/mnm$) and is isostructural with rutile. Each unit cell consists of two tin atoms and four oxygen atoms. The octahedral are sharing edges and form linear chains along the c-axis. n-type SnO_2 has been explored and widely used for transparent conductive electronic applications. In the past four decades, SnO_2 is the most extensively studied material, for gas-sensing applications, by using bulk, thick, and thin films to fabricate gas sensors. Most of the commercially available gas sensors of today are made mainly of SnO_2 in the form of thick films, thin films, or porous pellets. The well-known advantage of this material includes its low-cost, high sensitivities for different gaseous species and the ability to miniaturize them for integrating on micromachined substrates.

In the pure form SnO_2 is a semiconductor, with a bandgap of 3.6 eV, and its gas-sensing properties are recorded even in the non-stoichiometric form, SnO_x, with x values ranging between 1 and 2 [232,233]. In an ionic picture, Sn^{2+} has a $5s^2$ electron configuration. The natural growth faces of SnO_2 are mainly (110) and (100) surfaces. It is reported that the tin ions on the perfect surface of SnO_2 are all in the nominal Sn^{4+} state, as seen in the case of bulk SnO_2 material. The conduction and valence bands do not appear to be bent at this surface, that is, surface in a flat-band state [211]. Band bending does not take place as it is shown in Figure 4.1a. For this reason, the surface and the bulk materials exhibit similar resistivity values. The surface (110) of SnO_2 is thermodynamically most stable one. The interesting defect properties of this surface arise because of the bridging of oxygen ions, lying above the main surface plane. These oxygen ions can be removed easily either by heating the material, to higher temperatures, or by high energy particle bombardment. When they are removed, the two electrons left behind occupy orbitals, a mixture of $5s$ and $5p$ on surface tin ions converting them to Sn^{2+}.

It is a well-established fact that SnO_2 has an excellent potential as a gas-sensing material due to its high capacity to adsorb gaseous molecules and promote their surface reactions, usually their oxidation, while showing changes in surface conductivity. On the other hand, it is also well established that this material exhibits poor selectivity, known for its sensitivity to too many gaseous species, and high working temperatures required for better detection. Efforts to avoid these problems are partly successful but the challenge of producing high-performance material remains an open issue [234] till date. One of the most common ways to modify the characteristics of a material is introducing dopants. Selection of dopant is very crucial in this case.

Table A.2 list the different techniques used to synthesize nanostructured SnO_2 sensing material. They are aerosol technique [114,117], condensation from vapor phase [87], controlled solid–vapor process [60], deposition by novel low temperature aqueous chemistry technique [72], direct oxidation of tin at high temperature and *ex situ* mixing of CuO powders [122], hydrothermal treatment sol solution technique [103], hydrothermally treating α-stannic acid gel in ammonia solution at 200°C [106], laser-ablation technique with Nd:YAG laser with Sn target in an ambient of oxygen–argon mixture [98], MOCVD technique using tetramethyltin and oxygen at 500°C on different metal seeding regions [104], modified wet chemical route [118], PECVD technique using dibutyltin diacetate as a precursor and subsequent post-plasma treatment [101], prepared from supercritical fluid drying and calcination technique [107], sol–gel dip-coating technique [105,110–112,115,116,120], solvothermal treatment chemical process [35], synthesized with high-temperature chemical reaction method [100], thermal decomposition of Langmuir–Blodgett film precursors [96], thermal evaporation [99], thermal evaporation of oxide powders under controlled conditions without the presence of a catalyst [122], thermal evaporation of SnO powder without any catalyst [19], and vapor–solid growth methods [46].

Depending on the experimental conditions and on the involved processing parameters different nanostructures will take shape. Some of the reported nanostructures synthesized for this SnO_2 material are nanobelts [19,60,102], nanocrystalline films [111,115,117], nanoparticles [35,103–107], nanopowder [120], nanoribbons [122], nanorods [101], nanoscaled thin films [117], nanosized particles [107,118], nanowires [87,98–100], thin films of nanocrystals and nanopores [116], ultrathin films [96,112], and wide variety of other nanoshapes [72].

There are many factors that affect the gas-sensing properties of these materials. Beside the intrinsic semiconductor properties, several extrinsic factors, such as grain size, additives, microporous structure, and thickness of the gas-sensing entity, are also very important [103]. However, most of these reports are on conductivity type gas sensors, which are based on electrical conductivity change with varying gaseous concentration of a test gas in the ambient. Single crystalline SnO_2 nanobelts and nanowires are typically synthesized by using thermal evaporation or laser-ablation techniques. Alternatively, solution-phase SnO_2 synthesis has also been achieved using $SnCl_4$ precursor. In the case of SnO, SnO_2, and SnO/SnO_2 the equilibrium pressures of oxygen are quite close and the corresponding lines overlap with each other.

The use of SnO_2 as an active material for gas sensors is widely related to its combined use with catalytic metals. It has been claimed that their presence allows the following: decrease in sensor operation temperature, enhancement of sensitivity to different gases, and increase in response and recovery time for quickness [25]. From these catalytic additives, noble metals, mainly platinum and palladium, have been the most investigated from the point of view of the sensor response that they promote. It is expected that clusters are formed at the surface of SnO_2 and influence the surface reaction kinetics. These clusters will be in metallic or in oxidized forms depending on the noble metal deposited, the loading process, the interacting gas, and the sensor operation temperature.

SnO_2 is an oxide that is only slightly acidic; the basic characteristics of rare earth (RE) oxides may favor some catalytic aspects, such as adsorbing centers. Furthermore, it must also be considered that the introduction of RE cations (usually trivalent cations) influences the material's electronic distribution and the adsorption of oxygen species resulting in materials with suitable morphological characteristics such as high surface area and small particles. It is therefore expected that the concomitant use of noble metals and RE cations can produce some kind of synergetic effect, resulting in high performance of these oxide-based catalysts. Some experiments described in the literature point in this direction with results supporting those parameters discussed earlier.

Vayssieres [72] successfully demonstrated the growth of wide variety of SnO_2 nanostructures, by using a novel approach, on different substrates, which is based on thermo dynamical growth control concept. This is a low-temperature aqueous thin-film growth technique, done at 95°C, and was carried out in a large scale. Figure 7.42 shows the synthesized and purpose-built crystalline nanostructures of SnO_2 grown directly on ITO, single crystal silicon and on glass substrates. It is noticed that majority of these SnO_2 nanorods and nanowires predominantly grow along different crystal orientations. By careful consideration of symmetry and surface stability arguments, highly oriented bundles of *c*-elongated SnO_2 nanorods, with an aspect ratio of 1:10, with square cross section and (110) side faces, can be generated along with other anisotropic and porous nanostructures. The SEM pictures shown in this figure clearly explains the versatility of this technique to obtain different nanocrystalline forms of SnO_2 sensing material. The surface area of these nanostructures can be exploited for gas-sensing applications.

SnO_2 in the nanowire form has enormous potential to work as building blocks for nanoelectronics, and is also expected to offer superior chemical sensing performance due to the enhanced surface-to-volume ratio. Liu et al. [98] reported an efficient and reliable laser-ablation approach for large-scale synthesis of SnO_2 nanowires. Precise control over the nanowire diameters has been achieved by them by using monodispersed gold clusters as the catalyst. Figure 7.43a shows the SEM image of nanowires grown on Si–SiO_2 substrate. Figure 7.43b shows the XRD pattern of the synthesized SnO_2 nanowires. By using these nanowires *n*-type FET were made. Different transistor parameters

FIGURE 7.42
SEM images of purpose-built crystalline SnO$_2$ nanostructures grown directly onto various substrates of ITO, single crystal silicon, and glass by controlled aqueous growth technique at a temperature of 95°C. (Reprinted from *Comp. Rend. Chim.*, 9, Vayssieres, L., Advanced semiconductor nanostructures, 691–701, Copyright 2006, with permission from Elsevier SAS.)

were extracted for this nanowire FET structure. They have shown typical characteristics, such as ~−50 V threshold voltage and transistor ON/OFF ratio of ~10^3 at room temperature. Photoconduction properties on these wires showed a strong modulation of electrical conduction with UV illumination up to four orders of magnitude paving a new way to use them as polarized UV detectors

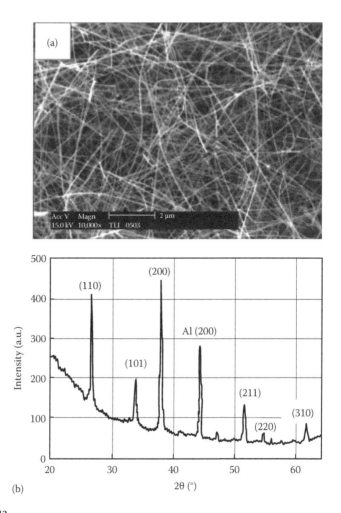

FIGURE 7.43

(a) SEM image of tin oxide nanowires grown on a Si–SiO$_2$ substrate. (b) XRD pattern of SnO$_2$ nanowires obtained on Si–SiO$_2$ substrate. (From Liu, Z., Zhang, D., Han, S., Li, C., Tang, T., Jin, W., Liu, X., Lei, B., and Zhou, C.: Laser ablation synthesis and electron transport studies of tin oxide nanowires. *Adv. Mater.* 2003. 15. 1754–1757. Copyright Wiley-VCH Verlag GmbH & Co. KGaA. Reproduced with permission.)

also. These types of nanowire FET devices are very attractive for the realization of micro-gas sensors. Its compatibility with silicon processing technology is an additional advantage for this type of devices. It is possible to exploit these FET structures for microsensor applications and the compatibility with silicon is an additional advantage for this material.

Kong and Li [122] reported highly sensitive SnO$_2$ nanoribbons, modified with CuO additive, that are sensitive to H$_2$S gas at room temperature. These nanofibers were synthesized by direct oxidization of tin powders at about 810°C. Small quantity of CuO was mixed to these nanowires to develop highly

FIGURE 7.44
SEM image of the synthesized SnO$_2$ nanoribbons. (Reprinted from *Sens. Actuators*, B105, Kong, X. and Li, Y., High sensitivity of CuO modified SnO$_2$ nanoribbons to H$_2$S at room temperature, 449–453, Copyright 2005, with permission from Elsevier B.V.)

sensitive and selective CuO–SnO$_2$ sensors for H$_2$S gas were fabricated. The sensitivity of these sensors was found to be very high and showed a switch like response to H$_2$S gas. It is expected that this type of combination could give better scope to develop newer nanodevices. Quick response and lower detection limit at 3 ppm is an additional advantage that is worth noting for these CuO–SnO$_2$ sensing elements. The response time increased with increasing of sensor operating temperature. It is reported that the recovery time, for these devices, on removal of H$_2$S is slow but improves with increasing temperature. These nanofibers are shown in Figure 7.44. Handling individual ribbons of these nanostructures is a real task.

Huang et al. [101] have demonstrated the synthesis of SnO$_2$ nanorods by using PECVD technique for initial depositions and subsequently plasma treatments were used to modify the microstructure of them. After the plasma treatment, uniform 1-D SnO$_2$ nanorods grown from the two-dimensional (2-D) films were observed in plasma-treated SnO$_2$ films. The nanorods were formed by a sputtering-redeposition mechanism and grew along their (101) preferential orientation. The optimal operating temperature of this plasma-treated SnO$_2$ thin film has decreased by 80°C whereas the gas sensitivity increased eightfold. The enhanced gas-sensing properties of these films may probably from the large surface-to-volume ratio of the nanorods' tiny grain size in the scale comparable to the space-charge length and its 1-D and 2-D hybrid microstructure.

Nanobelts of SnO$_2$ with a rectangular cross section, in a ribbon-like morphology, are very promising for gas-sensor applications. Since, the size of the depletion layer for tin oxide, due to oxygen ionosorption, penetrates 50 nm or more through the bulk, the belts are probably almost depleted of carriers as a pinched-off FET because this nanobelt thickness is typically less than 50 nm.

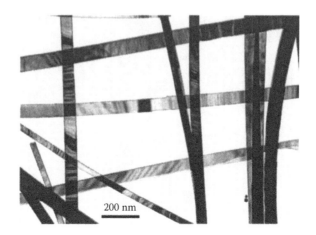

FIGURE 7.45

TEM image of as-synthesized SnO_2 nanobelts. (Reprinted with permission from Comini, E., Faglia, G., Sberveglieri, G., Pan, Z., and Wang, Z.L., Stable and highly sensitive gas sensors based on semiconducting oxide nanobelts, *Appl. Phys. Lett.*, 81, 1869–1871. Copyright 2002, American Institute of Physics.)

Figure 7.45 shows a TEM image of such SnO_2 nanobelts. Comini et al. [19] have reported that each nanobelt is a single crystal without the presence of any dislocations; its morphology and structure, such as growth direction and surface planes, are well defined and their surfaces are perfectly clean and atomically flat. The reported nanobelts have a rectangular-like cross section with typically an average width of ~200 nm, width-to-thickness ratios of 5–10 and lengths of up to a few millimeters. These are probably the longest single crystal nanostructures reported to any metal oxide. The contrast observed in the TEM image, of this figure, is mainly due to strain within the ribbons.

It is reported by Xu et al. [48] that the gas sensitivity of sintered SnO_2 elements with ultrafine particles, in the range of 4–27 nm, increased steeply when the average crystallite size, D, was below 10 nm, as shown in Figure 7.46a. Here, the effect of crystallite size on the resistance of nanoporous SnO_2 elements, for H_2 and CO is shown. Both the concentration values are at 800 ppm at an operating temperature of 300°C. In this figure, it is shown that when $D > 20$ nm the sensor's response signal R_a/R_g, where R_a is the resistance measured in clean air and R_g is the resistance measured in air mixed with the test gas, is nearly independent of the grain size. Below 20 nm, sensitivity increases with decreasing grain size, where below 10 nm this increase is found to be remarkable. Figure 7.46b shows the same data with $1/D$ for H_2 and CO. These results had a substantial impact on the design of metal-oxide gas sensors, leading to the development of various methods to produce sensors with a stable nanoporous microstructure. However, Rothschild and Komem [47] are at the opinion that the gas-sensing mechanism of nanocrystalline metal oxides is unclear and the effect of grain size on the gas sensitivity, in the limit of nanosized grains, requires further clarification.

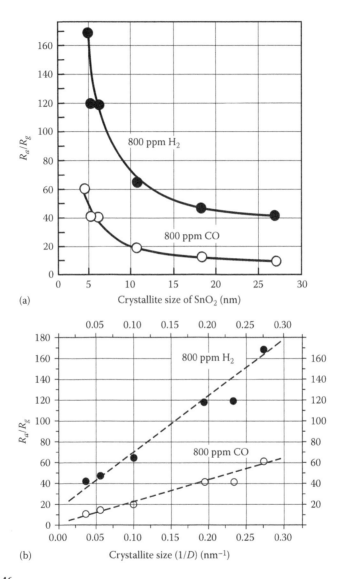

(a)

(b)

FIGURE 7.46

(a) The effect of crystallite size on resistance of nanoporous SnO_2 elements upon exposure to 800 ppm of H_2 and CO in air at an operating temperature of 300°C. (b) Same data as $1/D$ for H_2 and CO. (Reprinted with permissions from Rothschild, A. and Komem, Y., The effect of grain size on the sensitivity of nanocrystalline metal-oxide gas sensors, *J. Appl. Phys.*, 95, 6374–6380. Copyright 2004, American Institute of Physics; *Sens. Actuators*, B3, Xu, C., Tamaki, J., Miura, N., and Yamazoe, N., Grain size effects on gas sensitivity of porous SnO_2-based elements, 147–155, Copyright 1991, with permission from Elsevier.)

Single crystalline SnO$_2$ nanobelts of rutile structure were consistently synthesized, by Wang [60], by thermal evaporation of SnO$_2$ and SnO powders. HRTEM images revealed that these synthesized nanobelts are single crystalline in nature and were free from growth dislocations. Electron diffraction pattern evaluations indicated that the SnO$_2$ nanobelts grow along [101] crystalline orientation and it is enclosed by ±(010) and ±(10$\overline{1}$1) crystallographic facets. In addition to the normal rutile-structured SnO$_2$, it has been possible to form an orthorhombic super lattice-like structure. This structure can form in a thin nanowire, coexist with the normal rutile-structured SnO$_2$ in a sandwiched nanoribbon, or occur in the form of nanotubes. This result is distinct from that for bulk SnO$_2$, where pressures in excess of 150 kbar are generally required to form the orthorhombic form. Studying thermal and electrical transport, along the nanobelt, provides vital information about air ambience. By using photolithographically defined pad structures it is possible to use these nanobelts as a simple resistor exposed to test gas ambient. Figure 7.47a shows

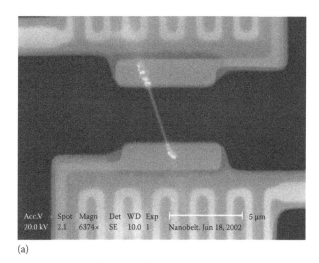

(a)

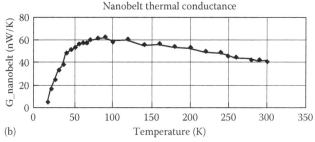

(b)

FIGURE 7.47

(a) Measurement of thermal transport through a single SnO$_2$ nanobelt. (b) Nanobelt thermal conductance variation at different temperatures. This type of structures is also useful for electrical transport studies. (Reproduced with permission of Annual Reviews Inc., from Wang, Z.L., *Annu. Rev. Phys. Chem.*, 55, 159, 2004. With permission through CCC Inc.)

one such measurement setup where nanobelts are used as a bridging resistor. Measurement of thermal and electrical transport through such single nanobelts provides the vital information of air ambience. Figure 7.47b shows the experimental observation for nanobelt thermal conduction variation at different temperature values. This is a new technique to develop small and compact micro-gas-sensor devices particularly for VOC, in vapor form, and general pollutant gaseous species detection. Handling such fine nanobelt structures is a real challenge for their development.

There exist a number of potential applications of these nanostructured materials; in particular the very long nanobelts have attracted attention for device applications. One problem of polycrystalline gas sensors is grain coarsening during operation that causes an alteration of the gas response. Gas sensors fabricated from single crystalline nanobelts could overcome this problem but still exhibit a high surface area to volume ratio essential for a strong gas response [235]. Furthermore, with further advancement of nanotechnology gas sensors based on single nanobelts may dominate in near future.

Chen et al. [236] reported ethanol sensing SnO_2 nanorods with extremely high sensitivity. These SnO_2 rods were synthesized through a hydrothermal route and were 3–12 nm diameter and lengths of 70–100 nm. Each nanorod has a sharp tip. The sensors fabricated from these nanorods exhibited excellent ethanol sensing properties with high sensitivity. This type of 1-D SnO_2 nanostructures is extensively applied to gas-sensing applications to detect NO_x, CO_x, H_2, C_2H_5OH, and H_2S gases. In case of C_2H_5OH gas, the sensitivity was linearly dependent on the concentration due to the small size of these nanorod structures. Other approaches like adding catalysts or doping techniques are yet to be identified. Such nanorods of SnO_2 are also reported without using any toxic organic chemicals during the synthesis [237]. The good sensing properties and easy fabrication of these nanorods is attracting many researchers for this field of gas sensors.

Baik et al. [103] synthesized SnO_2 nanoparticles by hydrothermally treated sol solution. Grain growth behavior of these films was systematically analyzed. Figure 7.48a shows an FE-SEM image of the thin film after sintering them at 600°C. The film consisted of a uniform stack of small grains with an average particle size less than 10 nm. The crystallite size of SnO_2 in the film was also determined from XRD analysis was of the order of 6–8 nm. Further, it was found that the grain growth is suppressed more in the thin film than in the powder. Figure 7.48b shows the electrical resistance of each device in air as a function of operating temperature (T). Electrical resistance of thin film shows temperature-dependent property. High-sensitivity region is identified and the range it covers. More details on temperature dependency are discussed in Figures 4.5 and 4.6 under Chapter 4 of this book.

The solvothermal reaction between metal alkoxides and benzyl alcohol enabled the synthesis of a variety of binary metal oxides under mild conditions [35]. The as-synthesized nanoparticles are crystalline in nature and exhibit uniform particle morphology with a narrow size distribution within one system. The SnO_2 sample, reported by Niederberger et al. [35],

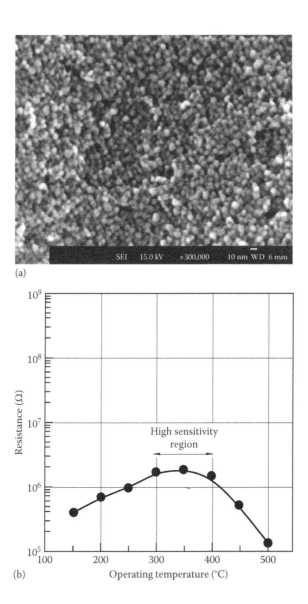

FIGURE 7.48
(a) High-magnification FE-SEM image of tin oxide film formed from hydrothermally treated sol solution. Calcination condition: 600°C, 3 h. (b) Electrical resistance of sensor device in air as a function of operating temperature. High-sensitivity region is identified for thin-film sensor derived from hydrothermally treated sol (sintered at 600°C). (Reprinted from *Sens. Actuators*, B63, Baik, N.S., Sakai, G., Miura, N., and Yamazoe, N., Hydrothermally treated sol solution of tin oxide for thin-film gas sensor, 74–79, Copyright 2000, with permission from Elsevier Science S.A.)

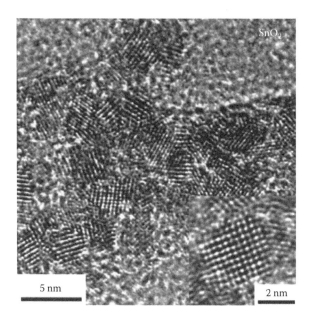

FIGURE 7.49
TEM micrograph of SnO_2 nanoparticles together with HRTEM image of single particle as inset. (Reprinted from *Prog. Solid State Chem.*, 33, Niederberger, M., Garnweitner, G., Pinna, N., and Neri, G., Non-aqueous routes to crystalline metal oxide nanoparticles: Formation mechanisms and applications, 59–70, Copyright 2005, with permission from Elsevier Ltd.)

consists of nanoparticles with an average size of 2–2.5 nm, as shown in Figure 7.49. The TEM micrograph shows the distribution of these nanoparticles. Despite the tiny particle size, the crystallinity of this SnO_2 material is really outstanding. The perfect distribution, within the grains, is shown in the figure inset.

Panchapakesan et al. [104] applied use of selected metal nanoparticles as seed layers for controlling the growth of microstructures of SnO_2 films on MEMS-based microhotplate devices with SiO_2 surfaces. High melting temperature metals such as iron, nickel, and cobalt, and relatively low melting temperature metals such as copper and gold were investigated as seed layers. Microstructural characterization has revealed the level of control that nanoparticle seeding can impart on the microstructure of SnO_2 films. Figure 7.50a through f presents the various SnO_2 microstructures produced by the different types of nanoparticle seeding. The same deposition conditions were maintained for SnO_2 deposition on different nanoparticle seed layers. The growth process is described as the onset of the nucleation of SnO_2 particles and the growth of SnO_2 particles into a continuous film. The micrographs show that while qualitative similarities exist between certain types of films, each type of seed metal produced a quantitatively unique SnO_2 morphology. Significant differences in microstructure were observed for these microhotplate-based films. Copper- and Ag-seeded SnO_2 films

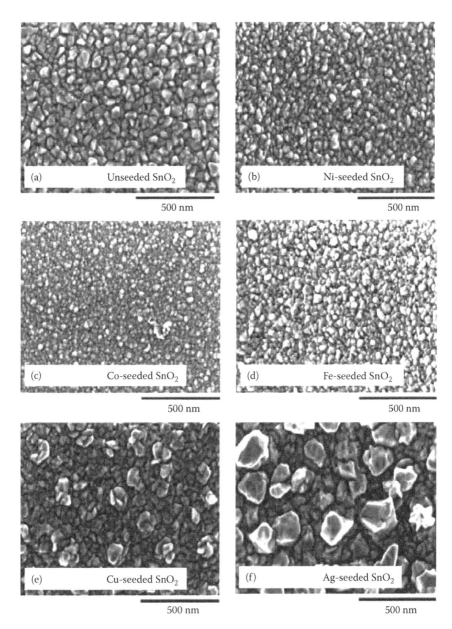

FIGURE 7.50
Effects of seeding on SnO_2 morphology: SEM images of (a) unseeded SnO_2, (b) nickel-seeded SnO_2, (c) cobalt-seeded SnO_2, (d) iron-seeded SnO_2, (e) copper-seeded SnO_2, and (f) silver-seeded SnO_2. (Reproduced with permission from Panchapakesan, B., DeVoe, D.L., Widmaier, M.R., Cavicchi, R., and Semancik, S., Nanoparticle engineering and control of tin oxide microstructures for chemical microsensor applications, *Nanotechnology*, 12, 336–349, 2001, Copyright IOP Publishing Ltd.)

showed morphologies that markedly differed from the other films. The qualitative morphological similarities between Fe-, Co-, and Ni-seeded SnO_2 films correlate well with the similar slopes of growth conductance. Although, the morphologies of Fe-, Ni-, and Co-seeded SnO_2 films were qualitatively similar, the grain sizes for these films differed. Using 90 ppm ethanol as a test case, seeding with high-temperature metals such as nickel, iron, and cobalt resulted in faster SnO_2 growth, smaller SnO_2 grain size and higher sensitivity. Furthermore, seeding was found to eliminate the effects of substrate morphology on the SnO_2 microstructure. The research team has also demonstrated the possibility of using multielement arrays to form a range of different types of sensor devices that could be used with suitable olfactory signal processing techniques in order to identify a variety of VOCs and pollutant gaseous species.

The nature of the metal seed strongly influences the final morphology of the grown SnO_2 film. This influence may be partly due to the different melting temperatures of the metal nanoparticles. Figure 7.51 is a plot of the average SnO_2 grain size versus the melting temperature of the seed-layer metal. It is seen that seed-layer metals with high-melting temperatures produced smaller

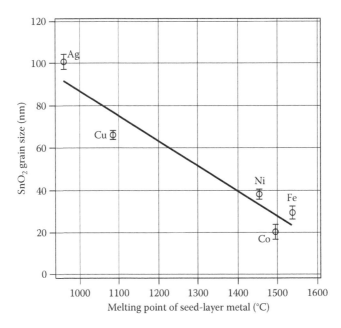

FIGURE 7.51
Correlation between nanoparticle seed layer melting temperature and SnO_2 grain size, illustrating that high temperature melting metals are good choices as seed layers for microstructural control of SnO_2 films. (From Panchapakesan, B., DeVoe, D.L., Widmaier, M.R., Cavicchi, R., and Semancik, S., Nanoparticle engineering and control of tin oxide microstructures for chemical microsensor applications, *Nanotechnology*, 12, 336–349, 2001, Copyright IOP Publishing Ltd. Reproduced with permission.)

SnO_2 grain structures compared to metals with low-melting temperature. From the observations, it is always advantageous to use high temperature melting metals and deposit them as seeding layers. This approach appears as a good choice for the purpose of selecting a particular grain size option. Correlation between nanoparticles seed layer melting temperatures and the grain size of SnO_2 is highlighted in this figure. Figure 7.52 presents the plot of the sensitivity ratio versus the SnO_2 grain size for the different types of nanoparticle seeded SnO_2 films from an array. In this case, each data point presents an average sensitivity values taken from six identical sensor devices from a single microhotplate array. The error bars represent the standard deviations from the average sensitivity values. This figure illustrates the increase in sensitivity as the grain diameter decreases for a test gas of 90 ppm ethanol. The increase in sensitivity is believed to be due to the increase in surface area of the grain boundary sensing sites due to the grain diameter decreasing. These results illustrate the benefits and advantages of nanoparticle SnO_2 for chemical gas-sensor applications using the seed-layer approach during the growth stage. This approach has demonstrated the suitability of fabricating devices with a precisely controlled grain size of sensing SnO_2 material.

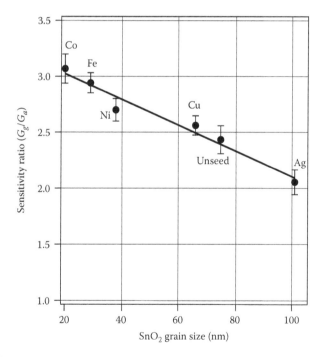

FIGURE 7.52
Gas sensitivity of nanoparticle-seeded SnO_2 films as a function of grain size in 90 ppm ethanol. (From Panchapakesan, B., DeVoe, D.L., Widmaier, M.R., Cavicchi, R., and Semancik, S., Nanoparticle engineering and control of tin oxide microstructures for chemical microsensor applications, *Nanotechnology*, 12, 336–349, 2001, Copyright IOP Publishing Ltd. Reproduced with permission.)

It is shown, by Han et al. [238], that the average crystal size of SnO_2 samples decreased with increasing the amount of vanadium doping. These V-doped SnO_2 nanocrystalline powder was synthesized by the co-precipitation method from vanadium(V) oxytrichloride and tin(IV) chloride. It is reported that the average crystal size has varied from 5.2 to 6.5 nm. The details of the variations are shown in Figure 7.53. The amount of vanadium doping was too small, in the present case, to affect XRD patterns, but it affects the crystal size and d-spacing value with (110) peak, for example, of XRD patterns. The XRD results show that V-doped SnO_2 nanocrystallites have rutile structure and the patterns are fairly broad. From the analysis, it is not clear whether the doped vanadium exists as V^{5+} ions, in SnO_2 crystal lattice, or vanadium oxides such as VO, VO_2, and V_2O_5. It is further reported that powder V-doped SnO_2 does not incorporate as V_2O_5–SnO_2 binary. It is further shown that the average crystal size of samples decreases with increasing amount of vanadium. The morphology of nanocrystalline 3 wt.% V-doped SnO_2 particles is spherical, and the distribution of crystal size is relatively uniform. This study shows the effect of vanadium doping to the crystal size. Presence of vanadium in SnO_2 lattice and its influence on the gas sensitivity parameters is not very clear. More results are expected in the near future.

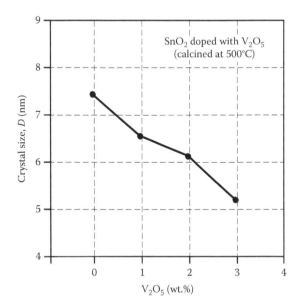

FIGURE 7.53

Effect of the amount of doped vanadium on the crystal size of SnO_2 samples calcined at 500°C. (Reprinted from *Sens. Actuators*, B66, Han, S.-D., Yang, H., Wang, L., and Kim, J.-W., Preparation and properties of vanadium-doped SnO_2 nanocrystallites, 112–115, Copyright 2000, with permission from Elsevier Science S.A.)

Pan et al. [31] have reported the gas-sensing properties of nanometer-sized SnO_2 doped with ruthenium and rhodium impurities. These doped nanoparticles were prepared by sol–gel method using inorganic salt as a precursor. The nanoparticle size was around 15 nm and the powder was annealed at 600°C for further analysis. The gas-sensing properties of these samples were carried out at different temperatures. The sensitivity of these doped sensors is shown in Figures 7.54 and 7.55. With reference to Figure 7.54, doping of 0.2 wt.% Ru catalyst in nano-SnO_2, the gas sensitivity has increased for petrol, C_4H_{10}, and H_2 gases. The same sensitivity for C_2H_5OH, CH_4, and CO is not apparent in comparison to other gases under study. The study indicates that the sensitivity for petrol and C_4H_{10} has increased because of the presence of ruthenium as an impurity. Since, the influence of temperature to the sensitivity is just in contrary, the petrol can be detected selectively in the temperature range between 200°C and 300°C where the gas sensitivity (S) is as high as 20 times and the selectivity coefficient

$$K = \frac{S_{petrol}}{S_{C_2H_5OH}}$$

is as high as 4 times. When the sensing temperature is higher than 300°C, the C_4H_{10} can be detected selectively where the gas sensitivity (S) is as high as 20 times and the selectivity coefficient also is as high as 4 times.

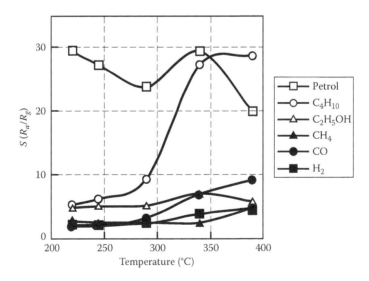

FIGURE 7.54
Gas sensitivity of nano-SnO_2-doped Ru varied with temperature. (Reprinted from *Sens. Actuators*, B66, Pan, Q., Xu, J., Dong, X., and Zhang, J., Gas-sensitive properties of nanometer-sized SnO_2, 237–239, Copyright 2000, with permission from Elsevier Science S.A.)

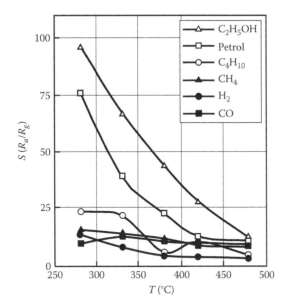

FIGURE 7.55
Gas sensitivity of nano-SnO_2-doped Rh varied with temperature. (Reprinted from *Sens. Actuators*, B66, Pan, Q., Xu, J., Dong, X., and Zhang, J., Gas-sensitive properties of nanometer-sized SnO_2, 237–239, Copyright 2000, with permission from Elsevier Science S.A.)

From Figure 7.55, it clearly shows that after doping 0.2 wt.% rhodium catalyst in nano-SnO_2, the gas sensitivities have increased largely for CH_4, CO, H_2, C_4H_{10}, C_2H_5OH, and petrol, and the gas sensor can detect the aforementioned reducing gases generally in the range between 260°C and 400°C, where the gas sensitivities are higher than 10. By selecting the impurities and incorporating them into SnO_2 it is possible to develop a specific sensing material for a particular gas. This present study has clearly demonstrated that gaseous species can be selectively detected by proper doping and by operating them at specific temperature ranges.

Cirera et al. [207] presented a new method, based on microwave treatment, to produce both doped and undoped nanosized SnO_2 gas-sensing powders. This is an inexpensive production method to synthesize material in just few minutes. Results of structural material characterization are introduced showing the suitability of derived nanopowders. Typical conventional heating treatments were employed for the stabilization of these nanopowders. It is reported that improved sensing properties were observed with palladium doping to SnO_2 and these sensors exhibited low cross-sensitivity behavior. Electrical measurements, after exposure to CO and NO_2 gases, indicate the feasibility of sensors improved with microwave derived SnO_2 nanopowders. As shown in Figure 7.56, conventionally treated powders exhibited similar results of maximum sensitivity temperature to CO exposure to those found in the published literature. Good results were obtained in 1000°C treated nanopowders for

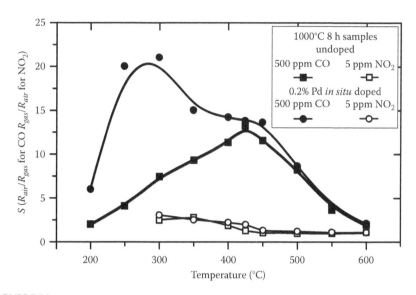

FIGURE 7.56
Sensitivity (ratio of resistance in air and resistance in gas) of undoped and *in situ* palladium doped SnO$_2$ sensors with 5 ppm NO$_2$ and 500 ppm CO. (Reprinted from *Sens. Actuators*, B64, Cirera, A., Vilà, A., Diéguez, A., Cabot, A., Cornet, A., and Morante, J.R., Microwave processing for the low cost, mass production of undoped and in situ catalytic doped nanosized SnO$_2$ gas sensor powders, 65–69, Copyright 2000, with permission from Elsevier Science S.A.)

undoped and palladium *in situ* doped sensors. The sensitivity of palladium *in situ* doped sensors show unusual behavior as their response was similar to undoped material. The figure also presents the sensitivity to the presence of NO$_2$ (5 ppm in synthetic air without humidity). These results show low cross-sensitivity of this gas with respect to CO, a point worth noting.

SnO$_2$ doped with rare earth material, such as samarium, was prepared by Wang et al. [239], by using hydrogen reduction method. Gas sensitivity of these doped SnO$_2$ elements was measured in 500 ppm n-C$_6$H$_{14}$, H$_2$, C$_2$H$_5$OH, C$_6$H$_6$, and CO in air at 250°C. The sensitivities of different samples to these gases are shown in Figure 7.57. The gas sensitivity of SnO$_{2-x}$(Sm) to the tested gases has shown a trend depending on the concentration of oxygen vacancies. When concentration of oxygen vacancies increases, the number of donor ions in SnO$_{2-x}$, the amount of active absorbed oxygen and specific surface area also increase. According to the theory of surface conductance of a semiconductor with a high x value, typically at a value of 0.5 and above, shows higher conductance because of the number of defects increases. This dopant is attractive because the operating temperature of the sensing element here in this case is about 250°C and it is easy to create such hot zones by using MEMS platforms defined either by bulk or surface micromachining techniques.

Optical gas-sensing properties of Ag- and Pt-doped SnO$_2$ films were studied by Yu et al. [240]. Ultrafine particle films, deposited by magnetron

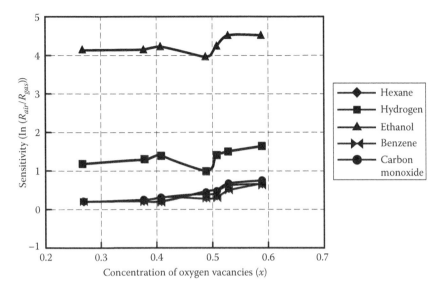

FIGURE 7.57
Sensitivity of Sm-doped SnO_{2-x} to 500 ppm tested gas at 250°C. (Reprinted from *Sens. Actuators*, B66, Wang, D., Jin, J., Xia, D., Ye, Q., and Long, J., The effect of oxygen vacancies concentration to the gas-sensing properties of tin dioxide-doped Sm, 260–262, Copyright 2000, with permission from Elsevier Science S.A.)

sputtering technique, have shown better gas-sensing optical spectrum property. Optical transmissivity of these films reduces with increasing alcohol concentration when they are exposed to in the ambient. This change in optical transmissivity tends to saturate at high concentration values. This type of sensing behavior is observed at room temperature. By measuring optical transmissivity it is easy to measure the concentration of alcohol in the ambient particularly at higher concentration values of alcohol. This novel approach finds application, to measure high gas concentration values, in the industrial segments and also in breathing tests. The relationship between transmissivity and concentration is shown in Figure 7.58, when wavelength is equal to 400 nm for four concentration values. The Ag-doped SnO_2 film has better optical gas-sensitive properties when gas concentration is lower than 10 vol.%. The Pt-doped film displays better optical gas-sensitive property when alcohol concentration is higher than 10 vol.% and in stipulated gas concentration region. Response time plays a key role in this type of sensors particularly for breathing applications. It is expected that a new series of optical devices may come with this concept in near future.

Catalytic metals such as platinum and palladium are often employed as surface dispersed additives on oxide gas sensors to improve sensitivity and selectivity. However, the addition of catalytic metals results in loss of sensitivity over long periods of time due to fouling or coking of catalytic metals, when exposed to certain gases and vapors [104]. The conventional processing

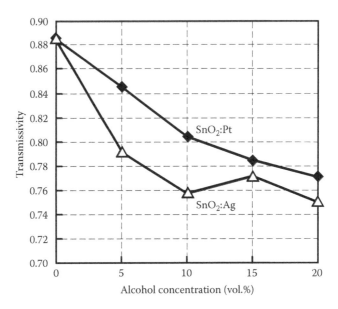

FIGURE 7.58
Relationship between transmissivity and alcohol concentration. (Reprinted from *Sens. Actuators*, B66, Yu, J.R., Huang, G.Z., and Yang, Y.J., The effect of gas concentration on optical gas sensing properties of Ag- and Pt-doped SnO_2 ultrafine particle films, 286–288, Copyright 2000, with permission from Elsevier Science S.A.)

methods for making nanostructured SnO_2 films either from solid-state or wet chemistry are not readily compatible with the batch fabrication techniques employed for fabricating the complementary metal-oxide-semiconductor (CMOS) compatible microhotplates, addressed by the team.

Chowdhuri et al. [123] reported enhanced catalytic activity of ultrathin CuO islands on SnO_2 films for sensing H_2S gas. It is pointed out by the group that ultrathin CuO in the form of dotted islands on SnO_2 film showed high sensitivity at an operating temperature, of the order of 150°C, with a fast response time, of the order of 14 s. This response time is smaller when compared with other similar sensors reported. The enhanced sensitivity and response speed of CuO dotted SnO_2 sensor in comparison to the other two sensors SnO_2 and SnO_2–CuO reveals the importance of the catalyst layer distribution on the film. Dissociated hydrogen available from the CuO–H_2S interaction spill over and the chemical interaction with the adsorbed oxygen on the SnO_2 surface is found to play a dominant role in this case.

Table A.16 elaborates the tin-oxide-based nanomaterials, nanostructures, different dopants, such as single metals, in combination with other materials, mixed with other metal oxides. Wide number of gases that are covered by this nano-SnO_2 includes acetone, acetylene, ammonia, benzene, butane, carbon dioxide, carbon monoxide, dimethyl methylphosphonate, ethyl alcohol, hexane, hydrogen, hydrogen sulfide, nitrogen oxides, oxygen, ozone, petrol (gasoline), and propane. Signal analysis is very crucial for this material and one may mislead with the total signal the sensor generates. The cross-sensitivity of this material is very high and the material responds to too many different gases. Sensor temperature, the environment in which the sensor is being operated, is one of the key factors along with the presence of impurities. One should know the background before quantifying the gaseous species for which the sensor is exposed to any ambient.

7.16 Titanium Oxides

Among the different oxides of titanium most thoroughly studied d^0 transition metal oxide TiO_2 is a thermodynamically favorable phase. The outer electronic arrangement for this metal is $3d^24s^2$ and the principal valence state correspondingly is +4. The other less stable states of titanium are +3 and +2. The sesquioxide, Ti_2O_3, is prepared by hydrogen reduction of TiO_2. The +2 oxide, TiO, can be obtained by high temperature reduction of the dioxide. TiO_2 occurs most commonly in a black or brown tetragonal form known as rutile. Less prominent naturally occurring forms are anatase and brookite with rhombohedral shape. Both rutile and anatase are white in pure form. The material exhibits predominantly (110) natural growth faces, but (011) and (010) are also common. The as-synthesized TiO_2 nanowires often appear in rutile ($a = 4.953\,\text{Å}$; $c = 2.958\,\text{Å}$) and anatase ($a = 3.78\,\text{Å}$; $c = 9.498\,\text{Å}$) crystal structures.

TiO_2 is an n-type semiconductor and is receiving increased attention in view of its advantageous properties such as: high dielectric constant, excellent optical transmittance, high refractive index, high chemical stability, and suitable energy bandgap. It can be processed into dielectric capacitor, optical coating layer, catalyst support, photocatalyst, and also for gas-sensing applications. Stoichiometric TiO_2 (110) surface is inert to room temperature chemisorption to any molecule. Surface defects are necessary for activation and to use this material for gas-sensing applications. When point defects are created, predominantly oxygen ion vacancies, they cause dramatic changes in their electronic structure of the material. Surface defects are created by ion or electron bombardment or by suitably heating the material to high temperatures [211].

This TiO_2 material has gained great importance in the field of gas sensing and many scientific groups are presently working on this material especially on its different nanostructures. Its properties are based on surface interactions with reducing or oxidizing gases, which, as a result, affects the conductivity of the film. With ultraviolet photon absorption, an electron-hole pair that can facilitate reduction and oxidation chemistry at the surface of the material is generated in the film. These redox reactions clean the surface by breaking down organic contaminants to form mainly CO_2 and H_2O molecules. TiO_2 films also demonstrate the ability to switch from hydrophobic to hydrophilic surfaces after irradiation with UV light, which, together with its photocatalytic properties, has resulted in demonstrating antifogging and self-cleaning capabilities [126]. TiO_2 films are used as the active layer on the recently commercialized self-cleaning windows.

Table A.2 lists the different techniques adapted to synthesize TiO_2 nanostructures. They are anodization of titanium foil in diluted HF in H_2O [30], anodization of titanium substrates in HF electrolytes [127], anodizing titanium sheet in a 1:7 CH_3COOH and 0.5% HF electrolyte solution [126], cathodic vacuum arc deposition technique [55], citrate–nitrate auto combustion method [128], films by spin coating [125], hydrolysis of aqueous $TiCl_4$ [129], hydrolysis of titanium tetra-isopropoxide [129], oxidation of titanium metal in aqueous 10% H_2O_2 at 80°C on Si–SiO_2 platform [133], preparation by electrospinning method [132], preparation by sol–gel technique [125], and by spraying precipitation technique by using titanium chloride and triethylamine (TEA) [54]. The nanostructures successfully synthesized are nanofibers [132], nanoparticles [129], nanopowders [128], nanotube arrays [126,127], nanotubes [30], spherical nanoparticles, nanohollow particles [54], sponge-like structures with nanoscale walls/wires [133], thin nanodimensioned films [125], and ultrathin films [55]. Nanostructures are normally produced from solution-phase growth methods including surfactant, sol–gel, electrospinning, hydrothermal, etc. Recently, thermal evaporation resulted single crystalline TiO_2 nanowires have been reported. This way titanium oxide provides stable nanostructures for different applications.

As a sensor TiO_2 (rutile in most cases) is used as humidity and pressure-sensitive material and as a gas sensing for gases such as H_2, O_2, and CO. In order to improve its sensitivity, a proper catalyst, such as Pt, is often required. Mei et al. [219] have used CeO_2 as a catalyst and reported that this material has electrochemical properties of storing or releasing oxygen, depending on the surrounding atmosphere variations, and is a suitable catalyst for the oxygen sensitivity of TiO_2. The area of grain boundaries of CeO_2 with TiO_2 would be greater as the size of TiO_2 particles became smaller, and the dispersion state of CeO_2 will be improved as well. It is further reported that the oxygen sensitivity of nano-CeO_2 coating nano-TiO_2 materials was better than that of nano-CeO_2 coating micro-TiO_2 materials.

A highly selective and easy to use alcohol sensor has always been in great demand in the biomedical, chemical and food industries; more recently in wine quality monitoring as an electronic nose. Furthermore, another possible application of this kind of sensor is the development of breath analyzer for health tests. Comini et al. [125] have demonstrated that nanocrystalline TiO_2 thin films doped with niobium and platinum, prepared by the sol–gel technique, were found to be anatase-phase dominated with grain sizes in a range of 20–50 nm. The Rutherford back scattering (RBS) and XPS analysis indicated that these films are also stoichiometric in nature. Figure 7.59a and b shows the SEM pictures of undoped and doped TiO_2 films. The undoped films exhibited porous

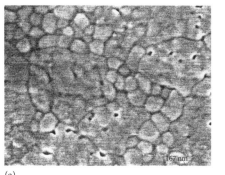
(a)

(b)

FIGURE 7.59
SEM pictures of (a) undoped titanium oxide and (b) doped titanium oxide films grown by sol–gel technique. Both the films were annealed at 800°C. (Reprinted from *Sens. Actuators*, B64, Comini, E., Faglia, G., Sberveglieri, G., Li, Y.X., Wlodarski, W., and Ghantasala, M.K., Sensitivity enhancement towards ethanol and methanol of TiO_2 films doped with Pt and Nb, 169–174, Copyright 2000, with permission from Elsevier Science S.A.)

texture where as the doped films are slightly nonporous. Both the films were annealed at 800°C. Experimental results of alcohol sensing of these, high temperature annealed, thin films indicate that the optimal operating temperature of the sensors ranges between 300°C and 400°C. The films have shown enhanced sensitivity for ethanol and methanol vapors compared to undoped films. The sensor response to 500 ppm ethanol was fast and as high as 2370% for TiO_2 doped with 1% Nb and also 0.5% Pt. The response and recovery dynamics to ethanol are particularly promising for applications in food analysis, wine identification, electronic noses, and breath analyzers [125]. These films are reported to be insensitive to other interfering gases such as CO, NO_2, and C_6H_6 but in combination with other metal oxides they do respond to CO gas that suits well with microcircuit fabrication technology.

Mor et al. [126] have reported a H_2 sensor based on TiO_2-nanotube arrays. The tubes were made by anodization of titanium sheet in an electrolyte, at 10 V potential, consisting of CH_3COOH and HF, with platinum counter electrode. Nanotubes of 22 nm diameter and wall thickness of 13.5 nm were obtained in this process. The details and the morphology of these tubes are shown in Figure 7.60a. This picture is shown from the top. The arrangement of all the tubes resembles as continuous netting. Figure 7.60b shows the cross-sectional view. The as-prepared amorphous titania nanotubes were subsequently annealed at 500°C for 6 h in oxygen ambient and this resulted crystalline titania having both anatase and rutile phases. The anatase phase is optimal for photocatalytic properties, while the rutile phase provides with its H_2 sensing capabilities. After crystallization, the TiO_2 nanotube film was sputter-coated with a thin, discontinuous palladium layer and this promotes catalytic dissociation of H_2 molecules, the ions of

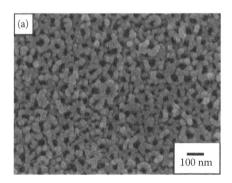
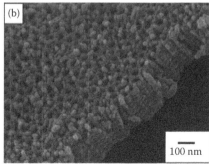

FIGURE 7.60

FE-SEM images of nanotube arrays prepared by using anodization technique at 10 V potential: (a) view from top and (b) cross-sectional view. (Reproduced from Mor, G.K., Carvalho, M.A., Varghese, O.K., Pishko, M.V., and Grimes, C.A., A room-temperature TiO_2-nanotube hydrogen sensor able to self-clean photoactively from environmental contamination, *J. Mater. Res.*, 19, 628–634, 2004, with permission from Materials Research Society and Cambridge University Press.)

which are then adsorbed on the surface of the TiO_2 nanotubes for sensing. At room temperature these sensors showed three orders of magnitude in their electrical resistance, for 1000 ppm H_2, which explains the capabilities of these nanosensors.

Varghese et al. [30] reported H_2 sensing with nanotubes having a diameter of 46 nm reaching a length of 400 nm. The surface morphology of these nanotube arrays were prepared using an anodization potential of 20 V and annealed at 500°C for 6 h in pure oxygen ambient. The surface morphology of these tubes is shown in Figure 7.61a at high magnification and Figure 7.61b at low magnification. The results indicate the reproducibility of nanotubes on the entire sample used for the experiment. Figure 7.61c shows another sample, anodized at 12 V, at higher magnification. It is observed that these nanotubes are uniform over the entire surface. The nanotubes annealed in pure oxygen ambient were found to be structurally stable to the temperatures of the order of 580°C. Above this temperature, protrusions were seen coming out through these nanotubes, an effect that spreads fast with increasing temperature. These protrusions, which were due to the oxidation of titanium substrate, make the entire nanotube structures to collapse and the network loses the physical shape.

The formation of titanium oxide nanotube arrays on titanium substrates was investigated by Zhao et al. [127] using HF electrolytes. Under optimized electrolyte and oxidation conditions, well-ordered nanotubes of titania were reported by them. These experiments were carried out at room temperature where solutions were stirred continuously using a magnetic stirrer. Microstructure details of these nanotubes are shown in Figure 7.62. Figure 7.62a shows the titanium metal microstructure that was polished mechanically (Figure 7.62b), the titanium metals after being anodized at 20 V in HF

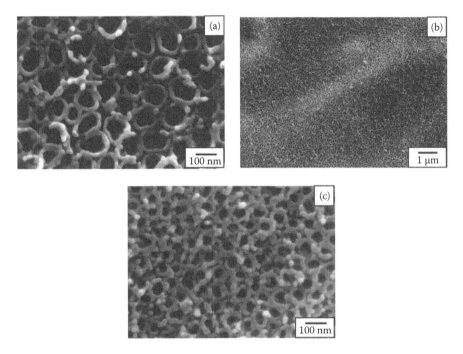

FIGURE 7.61
The surface morphology of the titania nanotubes after annealing at 500°C: (a) images at high magnification, (b) low-magnification images of a 20 V sample, and (c) a high-magnification image of a 12 V sample. (Reprinted from *Sens. Actuators*, B93, Varghese, O.K., Gong, D., Paulose, M., Ong, K.G., and Grimes, C.A., Hydrogen sensing using titania nanotubes, 338–344, Copyright 2003, with permission from Elsevier Science B.V.)

solution, (Figure 7.62c), titanium metals after being anodized at 20 V and then etched in 1 wt.% HF solution, and (Figure 7.62d) titanium metals after being anodized at 20 V for 30 min in 1 wt.% HF solution. The inset in Figure 7.62d shows the tubes at higher magnification. Topologies of these anodized titanium changed remarkably along with the changing of applied voltages, electrolyte concentration and oxidation time durations. Electrochemical determination and SEM studies indicated that the nanotubes were formed due to the competition of titania formation and dissolution under the assistance of applied electric field. A possible growth mechanism has also been presented by them. In contrast to porous alumina membranes, a different morphology was obtained in the present study. This type of nanotube arrays of titania are very interesting and expected to play an important role in gas-sensing applications because of their surface conditions. These features produce a large surface area and offer better sensitivity parameters, a prime requirement for the gas-sensing elements and devices based on them.

Fabrication of micro-TiO_2 hollow particles was done by Chou et al. [54] using titanium chloride precursor solution to interact with precipitation

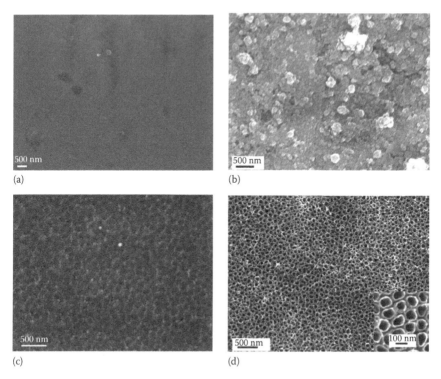

FIGURE 7.62
Titanium microstructure after different treatments: (a) titanium metals after being polished mechanically, (b) titanium metals after being anodized at 20 V above the HF solution, (c) titanium metals after being anodized at 20 V and then etched in 1 wt.% HF solution, and (d) titanium metals after being anodized at 20 V for 30 min in 1 wt.% HF solution. (Reprinted from *Solid State Commun.*, 134, Zhao, J., Wang, X., Chen, R., and Li, L., Fabrication of titanium oxide nanotube arrays by anodic oxidation, 705–710, Copyright 2005, with permission from Elsevier Ltd.)

agent TEA, which has hydrophilic–lipophilic property. The precipitation can be induced by acid–base neutralization while titanium chloride solution contact with TEA. Titanium precursor solution was prepared by mixing 20 mL titanium chloride with 60 mL of water. The droplets were sprayed and controlled at a relatively low falling velocity to impact with TEA liquid, no hole or concavities on the particle was found as shown in Figure 7.63a. The figure shows the SEM morphology of micro-titanium-oxide hollow particles with spherical features. After careful removal of TEA and vaporization of inside water the particles became hollow inside, similar with single hole particles because the formation of rigid shells. As shown in Figure 7.63b, most of the particles having a hole resulting from the sprayed droplet of titanium chloride solution controlled at high momentum to impact with the TEA liquid. Due to the fast agglomeration occurring on the boundary of droplet surface, titanium chloride

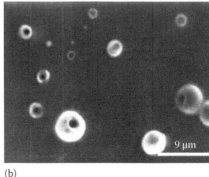

(a)   (b)

FIGURE 7.63
SEM morphology of micro-titanium-oxide hollow particle formed by spaying precipitation method with hydrophilic–lipophilic precipitation agents. (a) Spherical TiO$_2$ particles with hollow structures inside prepared by using titanium chloride–TEA system and the particles formed by controlling the sprayed droplets at low falling velocity. (b) Most of the micro-TiO$_2$ particles having a single hole prepared by using titanium chloride–TEA system and controlling the sprayed precursor droplets at relatively high falling velocity. (Reprinted from *Mater. Sci. Eng.*, A359, Chou, T.-C., Ling, T.-R., Yang, M.-C., and Liu, C.-C., Micro and nano scale metal oxide hollow particles produced by spray precipitation in a liquid–liquid system, 24–30, Copyright 2003, with permission from Elsevier B.V.)

precursor solution was solidified rapidly to form shell particles and the shell is too rigid to make the hole. These droplets offer more surface area for reactions.

Viswanathamurthi et al. [132] prepared TiO$_2$ fibers, doped with ruthenium, by electrospinning technique. This technique has been found to be unique and cost effective approach for fabricating large surface area membranes for a variety of applications. Electrospinning is a process by which high static voltages are used to produce an interconnected membrane like web of small fibers, with the fiber diameter in the range of 50–1000nm. Figure 7.64 shows the morphology of titania fibers with different content of ruthenium calcined at 600°C. The fibers appear no longer straight when the ruthenium content is 0.1 g and some of the fibers are broken if the ruthenium content is further increased. The exact reasons are not readily available.

Recently, Ishibashi et al. [249] investigated the formation of titanium oxide nanotubes through the anodization of metallic titanium, using ethanol and perchloric acid. They have demonstrated that titanium oxide nanotubes of more than 10 μm in length can be synthesized in a rather short time interval, ranging several minutes, through homogeneous anodization of a metallic titanium sheet, and also nanotubes with different tube diameters, by changing the anodization conditions, such as electrolyte composition and anodic potential values. Using this technique, dye-sensitized solar cells (DSSCs) were fabricated. These DSSCs are simple in structure but exhibit relatively high conversion efficiency comparable to that of amorphous silicon solar

FIGURE 7.64
SEM images of ruthenium-doped titanium dioxide fibers calcined at 600°C: (a) 0.1 g ruthenium, (b) 0.125 g ruthenium, and (c) 0.15 g ruthenium. (Reprinted from *Inorg. Chem. Commun.*, 7, Viswanathamurthi, P., Bhattarai, N., Kim, C.K., Kim, H.Y., and Lee, D.R., Ruthenium doped TiO$_2$ fibers by electrospinning, 679–682, Copyright 2004, with permission from Elsevier B.V.)

cells. Figure 7.65a and b show the FE-SEM image of these nanotubes fabricated through different anodization conditions. The picture in Figure 7.65a depicts a large bunch of nanotubes collectively put together. Inset shows the end portion of one such nanotube with an opening roughly equal to 100 nm outer diameter and 50 nm internal diameters. Similarly Figure 7.65b shows a bunch of fabricated nanotubes.

Taurino et al. [40] prepared thin nanostructured film of TiO$_2$, for VOC sensing applications, by using a new technique called seeded supersonic beam deposition. The advantage of this technique is the possibility to prepare nanostructured materials at room temperature and the possibility to control the grain dimensions during the deposition process. Based on this technique, a microsensor array of TiO$_2$-based sensitive films was used for the classification of three different alcohols with different concentration levels and at different working temperatures. This new technique allowed deposition of thin films with nanocrystalline properties and the film properties were tuned by varying the deposition parameters. In this way, the obtained films were utilized in order to implement a sensor array. An unsupervised linear method (Principal Component Analysis, PCA) as pattern recognition technique was used for evaluation of the alcohols.

The control of air-to-fuel (A/F) in combustion processes is one of the most important applications of oxygen sensors in automobile industry. A well-known example is the feedback control of A/F ratio of automobile engine exhaust gases in order to improve the fuel economy efficiency and to reduce the harmful

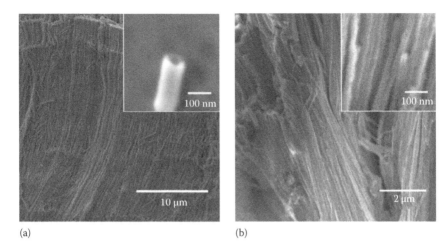

(a) (b)

FIGURE 7.65
FE-SEM images for anodic titanium oxide nanotubes fabricated through anodization under (a) condition A (ethanol and perchloric acid) and (b) condition B (perchloric acid), respectively. (Reproduced from Ishibashi, K.-I., Yamaguchi, R.-T., Kimura, Y., and Niwano, M., Fabrication of titanium oxide nanotubes by rapid and homogeneous anodization in perchloric acid/ethanol mixture, *J. Electrochem. Soc.*, 155, K10–K14, 2008, by permission of ECS—The Electrochemical Society.)

emission of gases such as CO, NO_x and different hydrocarbon exhausts [250]. Compared to the other oxygen gas sensors, the semiconducting oxide-based oxygen sensors exhibit many advantages for this specific application. In its rutile phase, TiO_2 is stable above 800°C and is employed as an oxygen sensor in such high temperature environments. In this respect, sensors based on TiO_2 have been most intensively investigated and found to be suitable for precise control of the A/F ratio operating near the lambda (λ) point. An important feature of this sensor is that the resistance changes significantly around this λ point.

Nanoparticle TiO_2 thin films appear to be promising candidates for alcohol sensing. Thin films prepared by using the sol–gel method offer certain advantages, such as the ability to grow the films over large area substrates, low processing temperature, very low cost of production facilities and a range of flexible chemical components [29]. Doping with both niobium and platinum enhances the response toward ethanol, while methanol is better detected by samples doped with niobium alone. Furthermore, niobium doping introduces a pronounced peak, in the response, at a temperature of 400°C. As far as propanol is concerned, it seems only a huge amount of platinum (equal to 0.5%) can catalyze its detection.

The sensor responses were calculated as the ratio R_{air}/R_{gas}, where R_{air} and R_{gas} are the electrical resistances of the sample in dry air, and mixture of the gas and dry air, respectively. It was found that the operating temperature of the sensor has a large influence on the gas response properties and dynamic behavior of the responses themselves. Figure 7.66a through c show the responses of the

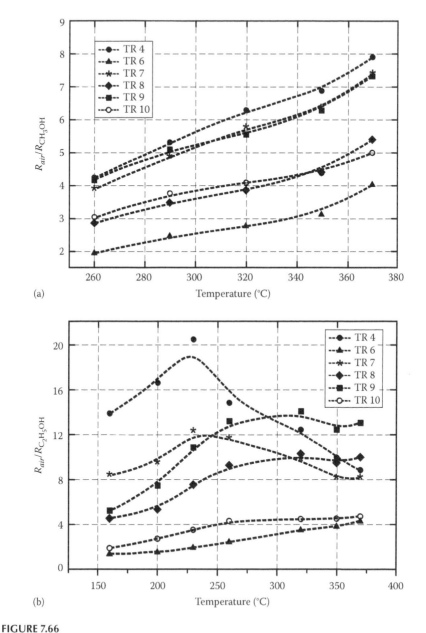

(a)

(b)

FIGURE 7.66
Titanium dioxide–based different nanocrystalline thin-film sensor responses at different working temperatures for three alcohols: (a) methanol, (b) ethanol and

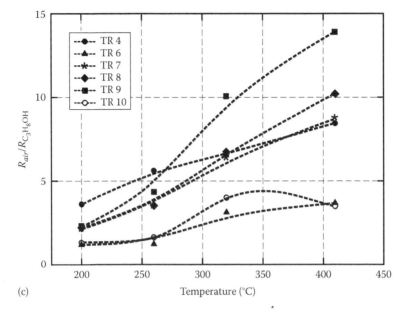

(c)

FIGURE 7.66 (continued)
(c) propanol. (Reprinted from *Sens. Actuators*, B92, Taurino, A.M., Capone, S., Siciliano, P., Toccoli, T., Boschetti, A., Guerini, L., and Iannotta, S., Nanostructured TiO₂ thin films prepared by supersonic beams and their application in a sensor array for the discrimination of VOC, 292–302, Copyright 2003, with permission from Elsevier Science B.V.)

array as a function of sensor operating temperature [40]. Here TR4, TR6, TR7, TR8, TR9, and TR10 are different investigated TiO_2 samples. The variation of the electrical resistance is different for the examined sensors, depending on the nature of each sensor and the alcohol under investigation. In fact, in the case of methanol and propanol, all the sensors have shown better responses in the highest temperature range, that is, between 370°C and 410°C. On the contrary, in the case of ethanol, the response curve goes through a maximum, around 250°C, and shows lower response values at subsequent temperatures. From these results, it is clearly evident that the sensors of the array, despite a better response at a particular working temperature that makes it possible to discriminate a specific gas, are not selective enough.

TiO_2 coatings are extensively used as optical materials, photocatalysts for photodegradation of organic pollutants in aqueous environments, and also as a component of DSSCs. TiO_2 xerogels have been prepared for preparation of microporous ceramic membranes. The thin films of these oxides have found considerable interest for UV filter applications and as integrated gas and humidity sensors. However, TiO_2 also exhibits its sensitivity to various other gaseous species as well and its cross-sensitivity has become an issue for signal analysis.

A comparison of the variation in resistance of samples having pore diameters of 46 and 76 nm, for 1000 ppm H_2 at 290°C, are shown in Figure 7.67 [30].

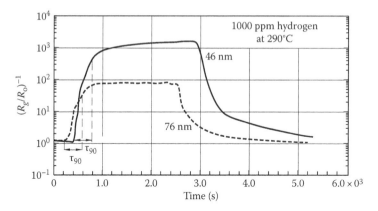

FIGURE 7.67
A comparison of the variation in resistance of samples having pore diameters of 46 and 76 nm, with time, upon exposure to 1000 ppm of H_2 at 290°C. (Reprinted from *Sens. Actuators*, B93, Varghese, O.K., Gong, D., Paulose, M., Ong, K.G., and Grimes, C.A., Hydrogen sensing using titania nanotubes, 338–344, Copyright 2003, with permission from Elsevier Science B.V.)

The results indicate that the nanotube sensors are capable of monitoring H_2 concentration levels ranging from 100 ppm to 4%. The nanotubes with smaller pore diameter (46 nm) showed higher sensitivity to H_2 when compared to larger pore diameter samples (76 nm). These sensors also showed high selectivity to H_2 when compared to CO, NH_3, and CO_2 gaseous species. Although the sensor was sensitive to high concentrations of oxygen, the response time was high and the sensor did not completely regain the original condition. It is believed that the H_2 sensitivity of the nanotubes is due to chemisorption onto the titania surfaces and they act as electron donors.

Niobium doped TiO_2 nanosized thick films were prepared by Ferroni et al. [251] for gas-sensing applications. The undoped TiO_2 powder exhibits large grain coarsening but niobium addition showed *n*-type behavior and it acted as a donor-dopant. The average grain size of Nb-TiO_2 grains is considerably lower, indicating that the driving force for grain growth is reduced by niobium addition. Microstructural characterization also showed that addition of this element inhibits grain growth and anatase-to-rutile transition. Differential thermal analysis (DTA) measurements confirmed that the role of niobium is to hinder structural modification in the powder. Niobium also affects the electrical properties of TiO_2. Depending on dopant addition and firing temperature, the samples differ in crystalline phase, grain size and turn out to be differently sensitive. Grain size is the parameter that determines the amplitude of the response of these sensors to CO gas. Dry and wet environmental monitoring is reported by Ruiz et al. [134] indicating that the humidity effects on gas sensor response toward CO and ethanol are less important in Nb doped samples than in undoped ones.

Nano-TiO_2 can greatly improve the antiageing properties of flame-retardant coating by its excellent ultraviolet-blocking power and nanoscale interpenetrating network formed by uniformly dispersed nanoparticles. This can enhance the resistance of the coating to moisture in natural weathering conditions [252]. This nanocoating can have a good expanding effect and fire-resistant property even after 500 h accelerated ageing tests. Francioso et al. [253] presented technology specific TiO_2 sensors for automotive application. Pd-doped TiO_2 devices onto ceramic substrates were deposited by sol–gel technique, with a heater on the backside, for commercial lambda (λ) probe. This type of device is expected to be highly useful in the automobile industry.

Details of titanium oxide nanomaterials are shown in Table A.17. These materials are sensitive to CO, C_2H_5OH, H_2, CH_3OH, nitrogen oxides, O_2, and for C_3H_7OH. Measurements are mainly based on the electrical conductivity variation in the films. Response times for these sensors are typically in the range of 1–3 min range. Nanostructures based on this material were synthesized using high-energy ballmilling, anodization, laser-assisted spray pyrolysis, RF sputtering, and standard chemical routes. Formation of TiO_2 nanotubes is a major attraction of this material. Sensor devices need slightly higher temperatures but well within the limits of thin-film-diaphragm-based MEMS structures. This advantage offers new openings to the field of gas sensors and related devices. It is expected that TiO_2-based nanostructures may dominate in near future for gas-sensing applications particularly those work on low power systems.

7.17 Tungsten Oxides

Chemically tungsten is relatively inert but forms compounds with most of the nonmetals. Tungsten compounds have been prepared, however, in which this element exhibits oxidation states ranging from 0 to +6. The states above +2, especially +6, are most common. In the +4, +5, and +6 states tungsten forms a variety of complexes. Its coordination number, referred as number of directly bound atoms, ranges from 4 to 9.

With oxygen tungsten forms different oxides such as WO, WO_2, WO_3, W_2O_3, and W_4O_3. Tungsten trioxide (WO_3) is a lemon-yellow solid and having a structure in which each tungsten atom is surrounded by six oxygen atoms in an octahedral array. Reaction of WO_3 with a stoichiometric amount of tungsten powder at 1000°C in a nitrogen ambient yields brown WO_2. The intermediate oxides, $W_{18}O_{49}$ and $W_{20}O_{58}$, are generally formed as a mixture, called blue oxide (with a typical composition equal to W_4O_{11}) upon heating WO_3, at about 800°C–1000°C temperature range, in a vacuum or in the presence of hydrogen gas. Both N_2O and NO oxidize WO_2 to blue oxide at 500°C, while NO_2 oxidizes WO_2 to WO_3 below 300°C.

Berger et al. [255] gave a detailed account of tungsten-oxide thin films, obtained by electron-beam deposition and subsequent annealing, in the temperature ranges of 350°C–800°C for varying time intervals. The changes of phase composition and the microstructure in dependence of the annealing conditions were described in detail. It is pointed out that the process of realignment of crystal structures, during solid-phase transformation, lead not only to the growth of new crystallites, with a preferential orientation, but also to a change in the direction of preferred growth with increasing annealing temperature and time durations. Various phase formations and structures are clearly brought out for the tungsten oxides.

WO_3 films are reported to have promising electrical properties for gas-sensing applications and they show a typical *n*-type conducting behavior. This particular phase shows a bandgap of 2.6 eV and has monoclinic crystal structure. These films are particularly attractive because they show a high catalytic behavior both in oxidation and reduction reactions on their surface. In addition, this oxide has many outstanding properties, making it a promising material in the fields of gas sensors, electrochromic application, photoelectrochromic "smart" windows, solar energy conversion, and electrocatalyst in fuel cells [151].

Recently, many efforts have been devoted to the morphology and size control of tungsten oxides due to their technological importance. As a result, various chemical or physical methods have been developed successfully to synthesize nanostructured tungsten oxides with special morphologies such as nanofibers spheres, nanotubes, nanowires, nanofibers, and nanorods. In addition to these features, stoichiometric and non-stoichiometric nanoparticles of tungsten oxides have been fabricated by electrodeposition method. This is achieved either by heating a simple tungsten filament or by the decomposition of pyrosol. Li et al. [151] reported the preparation of stoichiometric WO_3 nanoparticles with cuboid shape. These size-controllable growths of WO_3 nanocuboids are expected to facilitate understanding the relationship between the dimension and the sensing properties of these oxide materials.

Tungsten oxide nanomaterials are synthesized by a variety of other techniques, among which gas phase synthesis is of particular interest owing to its versatility, industrial viability, and many of its fundamental aspects. These materials find many uses in technology and give hopes for environmentally benign and resource-lean manufacturing. Many nanomaterials, particularly in thin-film form, are of interest for applications related to energy and the environment. Focusing only at the most recent work, it is observed that WO_3 can be employed for detecting hazardous pollutant gases such as H_2S and NO_x. Other gaseous species for example: various alcohol, different hydrocarbons, CO, and NH_3 are few other examples detected by WO_3 sensors as well.

Table A.2 lists the different techniques used for synthesizing these tungsten oxide nanostructures. They are advanced reactive gas deposition [159], advanced reactive gas evaporation technique [143], aqueous sols of tungsten oxide using ion-exchange method [148,149], catalytically modified chemical route with copper and vanadium additives [144], chemical processes and routes [135,158], controlled hydrolysis of WCl_6 in a coupling solvent [152], conventional thermal evaporation on ITO without any catalyst [139], deposition by supersaturated vapor nucleation onto carbon-coated copper grid [49,50], electrochemical etching of tungsten wires and subsequently heating in argon atmosphere at high temperatures [156], flat tungsten plates reduced with hydrogen at 700°C and later heated in argon atmosphere [156], hard template method [64], heating WS_2 in the presence of oxygen on surface etched tungsten foil [138], hot-wire CVD technique [94], infrared irradiation of tungsten foils under vacuum on tantalum substrates [137], one-step hydrothermal process [151], reactive RF magnetron sputtering technique [146], reactive thermal evaporation method [147], RF sputtering (with interruptions) [140], sol–gel technique using WCl_6 [150], solution drop coating [136], and thermal evaporation of tungsten at higher temperatures in the presence of oxygen [154]. Each process has its uniqueness. The nanostructures successfully synthesized are hollow fiber [138], hollow nanospheres [152], nanocrystallites [146–150], nanocuboids [151], nanofilms [140], nanoparticles [49,50,64,143,158,159], nanopowders [143,144], nanorod films [136], nanorods [94], nanotubes [137], nanowires [135,139,156], and nanowire networks [154].

Hoel et al. [49] deposited tungsten oxide (WO_3) nanoparticles by gas evaporation and subsequent deposition technique. By this method, WO_3 nanoparticles were deposited based on a simple principle: a vapor is created by evaporation of metal, or through sublimation, from a heated substrate material. This vapor is then cooled by collisions with an inert or reactive gas molecules in such a way that a supersaturated state is obtained from which nanoparticles are produced by nucleation and growth. These nucleated particles are subsequently deposited onto a substrate. If sufficient oxygen is introduced into the evaporation chamber at this stage, and the metal has any stable oxide form, either the metal atoms are oxidized in the vapor or an oxide layer is formed on the surface of the heated substrate. The latter situation takes place in case of tungsten, whose melting point is very high. Since WO_3 has a much lower melting temperature than tungsten metal, a sufficient amount of oxide can then be sublimated for WO_3 nanoparticle formation in this way. The details of these synthesized nanoparticles are shown in Figure 7.68. Figure 7.68a shows a TEM image of gas deposited nanoparticles, condensed from a vapor generated by heating solid tungsten in a flow of synthetic air. Carbon coated copper grid was used here for the condensation of these nanoparticles. A corresponding selected area diffraction pattern, of these nanoparticles, is shown in Figure 7.68b; it indicates that the nanoparticles have a tetragonal phase of WO_3. It is further proved, by Hoel et al. [49], that these nanostructured WO_3 sensors could detect the presence of a probing gas even at room temperature. As an example, 5 ppm of H_2S yielded a conductance change as large as 250 times. Doping of these WO_3 films with aluminum or gold metals increased the sensitivity, for H_2S gas, still further. It is reported that the maximum sensitivity was obtained for H_2S, N_2O, and CO test gases at 400, 525, and 700 K, respectively, for aluminum doped WO_3 nanoparticle sensors.

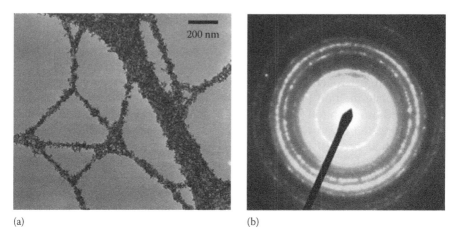

(a)   (b)

FIGURE 7.68

(a) TEM micrograph showing WO_3 nanoparticles deposited onto a carbon-coated copper grid. (b) Shows a selected area diffraction pattern from a WO_3 nanoparticle film. The ring pattern is consistent with the tetragonal phase of WO_3. (Reprinted from *Curr. Appl. Phys.*, 4, Hoel, A., Reyes, L.F., Heszler, P., Lantto, V., and Granqvist, C.G., Nanomaterials for environmental applications: Novel WO_3-based gas sensors made by advanced gas deposition, 547–553, Copyright 2004, with permission from Elsevier B.V.)

Depero et al. [254] deposited $W_{0.9}Ti_{0.1}$ thin films, instead of pure tungsten, by reactive sputtering technique. Annealing of these deposited films at 600°C resulted in a sensing oxide layer consisting with nanocrystallites of 30 nm in size. These films showed improved sensitivity to the NO_2 gaseous species. This feature is ascribed to the solution of titanium ions within the WO_3 lattice, which prevents the formation of coarse-grained structures and allows synthesis of a high surface-to-volume layer. Pure WO_3 segregation, as insulated crystallites, which do not significantly affect the sensing characteristics, suggests that future depositions should be carried out from a sputtering target richer in titanium than with pure tungsten. Annealing at 800°C produces a thin film of TiO_2 anatase with nanosized structure. Its high surface-to-volume ratio and nanogranularity envisage an exploitation of this material for NO_2 detection in the temperature, particularly in the range of 350°C–800°C.

Nanocrystalline porous tungsten oxide, in thin-film form, was synthesized by Wang et al. [150] by sol–gel method using tungsten hexachloride, WCl_6, precursor. This oxide film was formed on a highly polished alumina substrate by spin-coating technique. Figure 7.69a shows the surface morphology of the porous tungsten oxide films spin coated onto alumina substrates and calcined at 350°C. Similarly Figure 7.69b through d shows films at 450°C, 550°C and 650°C, respectively. The grains are very uniform in their size, only with the sample annealed at 650°C, but are distributed in a random way. Experimental results indicated that this film had high degree of sensitivity

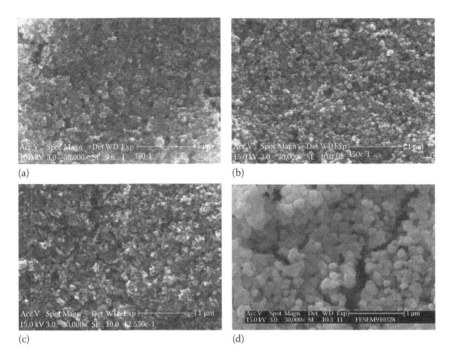

FIGURE 7.69
Surface morphologies of porous tungsten oxide thin films spin coated onto highly polished alumina substrates calcined at different temperatures: (a) 350°C, (b) 450°C, (c) 550°C, and (d) 650°C. (Reprinted from *Sens. Actuators*, B94, Wang, S.-H., Chou, T.-C., and Liu, C.-C., Nano-crystalline tungsten oxide NO_2 sensor, 343–351, Copyright 2003, with permission from Elsevier B.V.)

to low level NO_2 concentration in the range from 50 to 550 ppb with relatively fast response, typically of the order of 3 min, and recovery times. The sensitivity of the prepared tungsten oxide film depended on the surface structure, the grain size, and the geometrical heterogeneity of the films that were controlled mainly by the calcination temperature. The sensitivity of WO_3 to NO_2 also depended on the gas molecules NO_2 adsorbed form on the surface which was affected by the operating temperature. The optimal sensing condition was observed with the films using a tungsten oxide calcined at 550°C for 1 h and the operating temperature of the sensor at 300°C.

Palladium doped nanocrystalline WO_3 films were prepared by Hoel et al. [159]. Advanced reactive gas deposition technique was used to fabricate these nanocrystalline WO_3 films doped with palladium nanoparticles. Gas-sensing properties of these deposited films, on alumina substrates, with regard to H_2S and other types of gases such as CO, NO_2, SO_2, H_2, and HCHO were studied. The as-deposited films consisted of nanoparticles with sizes of approximately 5 nm for both WO_3 and Pd. The surface morphology of as-deposited and Pd-doped films, sintered at 600°C, is shown in Figure 7.70.

FIGURE 7.70

Scanning electron micrographs depicting surfaces of films of (a) as-deposited WO$_3$ and (b) WO$_3$: Pd0.5% sintered at 600°C. (Reprinted from *Sens. Actuators*, B105, Hoel, A., Reyes, L.F., Saukko, S., Heszler, P., Lantto, V., and Granqvist, C.G., Gas sensing with films of nanocrystalline WO$_3$ and Pd made by advanced reactive gas deposition, 283–289, Copyright 2005, with permission from Elsevier B.V.)

Figure 7.70a shows the surface morphology of as-deposited films and Figure 7.70b shows the films doped with 0.5% palladium sintered at 600°C. Upon sintering, the sensor materials have changed from tetragonal to monoclinic crystal structure for the WO$_3$ and from metallic palladium, via PdO, to PdO$_2$. The transition to PdO started at the temperature 200°C, and subsequently the palladium was fully oxidized to PdO$_2$ at about 600°C. These sensors exhibited room temperature sensitivity for all these gases referred earlier.

Jiménez et al. [157] have demonstrated that addition of impurities such as copper and vanadium improves the sensing properties of WO$_3$ for ammonia gas. These additives were introduced at different concentrations and the resulting material was annealed at different temperatures so as to achieve the optimal material for NH$_3$ detection. Sensors based on copper-catalyzed WO$_3$ presented the best sensor response to NH$_3$, with low interference of humidity, while gas sensors based on vanadium-catalyzed WO$_3$ had a sensor response to NH$_3$ highly dependent on humidity.

Indium metal doped nanoparticle thick films of WO$_3$ were prepared by Khatko et al. [158] by chemical route. Gas sensors were prepared using commercially available WO$_3$ nanopowders and powder mixtures with different concentrations of indium metal (1.5, 3.0, and 5.0 wt.%). Figure 7.71a and b shows the SEM image of pure films and Figure 7.71c shows the image of those indium-doped (3 wt.%) thick films screen printed on to alumina substrates. Here also the granules are randomly distributed. The morphology of these indium doped sensing layer did not depend on the metal concentration. These layers were full of nanoparticles with an average granule size of approximately 70 nm and had uniform granule distribution as shown in the figure. It was further reported that the surface morphology of the pure and indium doped WO$_3$ sensing layers is almost similar. There is just a small difference in the overall average granule size of these nanoparticles. Gas-sensing properties of these indium doped sensors to NO$_2$, CO, NH$_3$, and

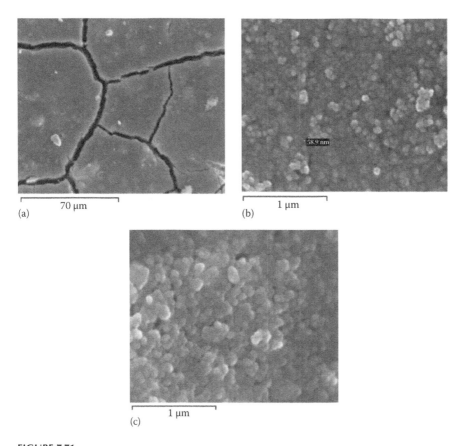

FIGURE 7.71
SEM images of WO_3 thick films: (a) and (b) pure films, (c) indium-doped WO_3 thick film with indium concentration of 5 wt.%, with an average granular size of 70 nm. (Reprinted from *Sens. Actuators*, B111–B112, Khatko, V., Llobet, E., Vilanova, X., Brezmes, J., Hubalek, J., Malysz, K., and Correig, X., Gas sensing properties of nanoparticle indium-doped WO_3 thick films, 45–51, Copyright 2005, with permission from Elsevier B.V.)

C_2H_5OH showed that the sensing layers began to respond even at room temperature. Such operations are quite useful for the development of sensors working at room temperature.

The studies of Ferroni et al. [204] gave the details of the titanium (10%)-tungsten (90%) films, deposited by RF reactive sputtering technique. These nanosized thin films were annealed by heating them in air at different temperatures. The morphological and structural characteristics were studied by using electron microscopy techniques. These studies on the microstructural properties of a W-Ti-O thin film by SEM and TEM analyses highlighted the features of the layer after annealing at either 600°C or 800°C. Annealing of these films at 600°C, as shown in Figure 7.72a, resulted in a sensing layer with enhanced sensitivity to NO_2 with respect to pure-WO_3 thin films. The fine granularity of the

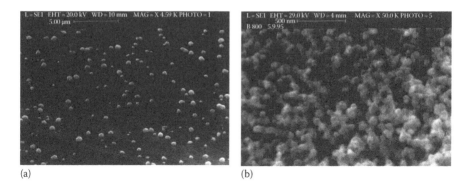

FIGURE 7.72
SEM microphotographs of the W-Ti-O films (a) annealed at 600°C and (b) annealed at 800°C. (Reprinted from *Sens. Actuators*, B44, Ferroni, M., Guidi, V., Martinelli, G., Nelli, P., and Sberveglieri, G., Gas-sensing applications of W-Ti-O-based nanosized thin films prepared by r.f. reactive sputtering, 499–502, Copyright 1997, with permission from Elsevier Science S.A.)

continuous layer determines its conducting properties and explains the high sensitivity of the material. This feature impedes formation of coarse-grained structures. Pure WO_3 segregation do not affect significantly the sensing characteristics, corroborates that depositions should be carried out from a sputtering target richer in titanium to take the advantage of the higher sensitivity factor. Annealing at 800°C produces a thin film of TiO_2 anatase with nanosized structure, as shown in Figure 7.72b, only partially covered with large WO_3 crystallites. SEM image shows that the film is not continuous here. The high surface-to-volume ratio and nanogranularity envisage an exploitation of this material. Ferroni et al. [204] have further concluded that reactive sputter deposition followed by annealing at 800°C of the sample is also a novel method to produce thin films of nanostructured TiO_2 anatase that are stable and is relatively easy to synthesize.

Dillon et al. [94] have demonstrated HWCVD technique to synthesize a variety of tungsten oxide nanomaterials. The HWCVD-generated material was a fine powder consisting of predominantly nanorods and nanotubes along with a few larger crystalline particles and a small amount of amorphous materials. Both bulk SWNT and MWNT were also identified in this synthesized powder material. These nanotubes were found to be crystalline rods with approximately 10–20 nm in diameter and 40–60 nm in length. Figure 7.73a displays a TEM image of these tungsten oxide nanorods and nanotubes. The crystalline nature and the perfection of one of the nanorods are clearly depicted in the HRTEM image, shown in Figure 7.73b. XRD analysis revealed that the crystalline phase of the tungsten oxide is WO_3. These types of nanostructures are expected to play an important role for gas-sensing application because of their large exposed surface area of these nanotube structures of these tungsten oxides.

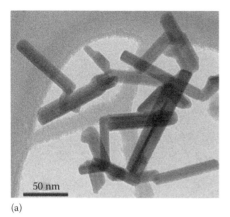

(a)   (b)

FIGURE 7.73
TEM images of (a) WO$_3$ nanorods, and (b) the same WO$_3$ nanorods at higher resolution revealing their crystallinity. (Reprinted from *Thin Solid Films*, 501, Dillon, A.C., Mahan, A.H., Deshpande, R., Alleman, J.L., Blackburn, J.L., Parillia, P.A., Heben, M.J., Engtrakul, C., Gilbert, K.E.H., Jones, K.M., To, R., Lee, S.-H., and Lehman, J.H., Hot-wire chemical vapor synthesis for a variety of nanomaterials with novel applications, 216–220, Copyright 2006, with permission from Elsevier B.V.)

WO$_3$ nanoparticle-based sensors were sensitive to H$_2$S gas, at room temperature, but the response times were of several minutes and recovery times were of several hours. This pollutant H$_2$S gas has drawn much attention because of its toxicity and its wider use in chemical laboratories and industries. H$_2$S gas also liberates in nature due to biological processes and also from in mines and petroleum fields. Rout et al. [135] have investigated the sensing characteristics of these WO$_3$ nanoparticles and nanoplatelets and of WO$_{2.72}$ nanowires toward H$_2$S in the 1–1000 ppm concentration range at working temperatures varying in the range of 40°C–250°C. Their study showed that WO$_{2.72}$ nanowires are good candidates for sensing H$_2$S in the 10–1000 ppm range at 250°C. Stoichiometric nanoparticulate WO$_3$ films show good response to 1 ppm of H$_2$S at 200°C. Active layers of pure and Pt-doped WO$_3$ films deposited by RF magnetron sputtering were able to sense 100 ppb of H$_2$S at 200°C. RF sputtered WO$_3$ films and films doped with platinum, gold, silver, titanium, SnO$_2$, ZnO, and ITO have shown improved response when they are doped with gold to H$_2$S gas. Figure 7.74a shows a FE-SEM image of these WO$_3$ nanoparticles, with the inset showing a TEM image and the SAED pattern. The SAED pattern indicates that these nanoparticles are single crystalline in nature. Figure 7.74b shows a FE-SEM image of WO$_3$ nanoplatelets with a TEM image as the inset. The TEM image reveals that the platelets are of 60 ± 20 nm long and 1–5 nm thick. It is further reported that the thickness of the WO$_3$ platelets are very thin and gets destroyed very fast by the scanning electron beam during the analysis. Figure 7.74c shows a TEM image of the WO$_{2.72}$ nanowires. The average diameter of these nanowires is in the range of 5–15 nm. The inset in Figure 7.74c shows a high-resolution

(a)

(b)

FIGURE 7.74
FE-SEM images of (a) tungsten oxide nanoparticles with the inset showing a TEM image and electron diffraction and (b) tungsten oxide nanoplatelets with the inset showing a TEM image.

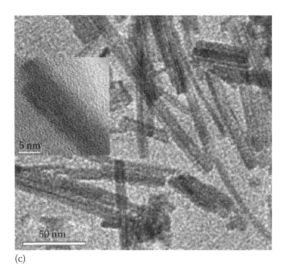

(c)

FIGURE 7.74 (continued)
(c) A TEM image of $WO_{2.72}$ nanowires with the inset showing a HRTEM image. (Reprinted from *Sens. Actuators*, B128, Rout, C.S., Hegde, M., and Rao, C.N.R., H_2S sensors based on tungsten oxide nanostructures, 488–493, Copyright 2008, with permission from Elsevier B.V.)

image of a nanowire. The single crystalline nature of this nanowire is seen from the HREM image, with a lattice spacing of 3.78 Å corresponding to that of the (010) crystalline planes.

Kim et al. [136] have demonstrated silicon MEMS-based miniaturized sensors by using highly crystalline tiny $WO_{2.72}$ nanorods, in thin-film form. The sensing layer was deposited by drop coating of $WO_{2.72}$ nanorod based solution. The sensors were fabricated on silicon wafers with a square membrane embedded with interdigitated electrodes and a platinum microheater. Cross-sectional view of the complete sensor structure is shown in Figure 7.75a. The microheater and detection electrodes are embedded with proper electrical isolation. Figure 7.75b shows the plane-view optical microscope image shows the interdigitated detection electrodes and the microheater. Figure 7.75c shows the SEM image of sensing layer. Gas-sensing measurements, on these fabricated sensors, were carried out in a small testing chamber. It is reported that these MEMS microsensors have shown high sensing ability for various reducing and oxidizing analytes even at room temperature. When exposed to reducing gases, the temperature-dependent response behavior was found to reverse from normal trend. This change over was observed at about 70°C. It was also felt that this type of unusual behavior may be due to unique structural features of these non-stoichiometric $WO_{2.72}$ nanorod films possessing high surface-to-volume ratio and active adsorption sites. These films probably have more favorable absorption sites due to its oxygen-deficient defect structure than the normal stoichiometric WO_3 with several

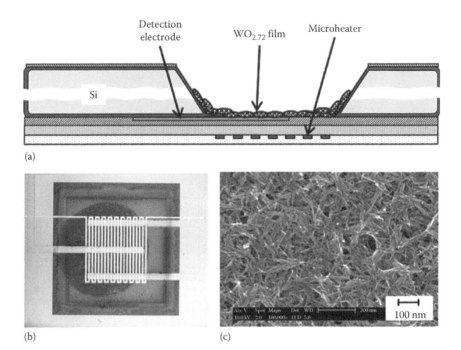

FIGURE 7.75

(a) Cross-sectional schematic diagram for well structure of a sensor substrate with a membrane-based hotplate, (b) plane-view optical microscope image of a fabricated sensor equipped with interdigitated detection electrodes, a microheater, and a sensing film, and (c) surface SEM image of the sensing layer deposited by drop coating of $WO_{2.72}$ nanorod based solution. (Reprinted with permission from Kim, Y.S., Ha, S.-C., Kim, K., Yang, H., Choi, S.-Y., Kim, Y.T., Park, J.T., Lee, C.H., Choi, J., Paek, J., and Lee, K., Room-temperature semiconductor gas sensor based on nonstoichiometric tungsten oxide nanorod film, *Appl. Phys. Lett.*, 86, 213105. Copyright 2005, American Institute of Physics.)

active sites. The facile vapor detection of $WO_{2.72}$ nanorod sensor at ambient temperatures might be successfully employed for the development of miniaturized sensing system thus fulfilling the requirement of low power consumption for battery operated devices. This type of devices with tungsten oxide is highly useful for low power field applications.

Three-dimensional tungsten oxide (WO_{3-x}) nanowire networks have been demonstrated, by Ponzoni et al. [154], as a high-surface area material for building ultrasensitive and highly selective gas sensors. These networks were prepared by thermal evaporation of tungsten metal, at a temperature in the range of 1400°C–1450°C, in the presence of oxygen. Figure 7.76a shows the low magnification SEM image of tungsten oxide nanowire network and Figure 7.76b shows the SEM image of 3-D nanowire networks exhibiting huge surface area highly suitable for gas-sensor applications. The inset in Figure 7.76a shows the EDS analysis of the tungsten oxide nanowire networks. In the present case gas sensors were fabricated by using drop coating technique

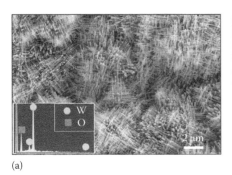

(a)

(b)

FIGURE 7.76
(a) Low-magnification SEM image of tungsten oxide nanowire network and (b) nanowire network at higher magnification. Inset in (a) is the EDS analysis of the tungsten oxide nanowire networks. (Reprinted with permission from Ponzoni, A., Comini, E., Sberveglieri, G., Zhou, J., Deng, S.Z., Xu, N.S., Ding, Y., and Wang, Z.L., Ultrasensitive and highly selective gas sensors using three-dimensional tungsten oxide nanowire networks, *Appl. Phys. Lett.*, 88, 203101. Copyright 2006, American Institute of Physics.)

on alumina substrates. Electrical conductivity measurements were carried out to detect the sensing gases such as NO_2 and H_2S by interdigitated structure. By utilizing the 3-D hierarchical structure of the networks, high sensitivity has been obtained toward NO_2, revealing the capability of the material to detect gas concentration levels as low as 50 ppb. The distinctive selectivity at different working temperatures was also observed for various gases. The results highlight that this particular technology can be adopted for the development of micro-gas sensors with performances suitable for practical applications. In particular, the capability to reveal NO_2 concentrations comparable with the threshold limit of outdoor application has been observed, together with low cross-sensitivity toward possible interfering gases such as H_2S, NH_3, and CO. The outdoor application, due to its low threshold limit, is considered as the frontier also for the thin- and thick-film technologies, typically adopted in gas-sensing field.

Highly sensitive hollow spheres of WO_3 were synthesized, by Li et al. [152], in solution phase, by controlled hydrolysis technique. These hollow spheres have tailored structures, exhibit low densities, high surface areas, and unique optical, electrical, and surface properties. It is expected that these materials will meet the needs of many scientists for different promising applications in the vast areas of miniature gas sensors, microchip reactors, and photonic crystals. Figure 7.77a is the low-magnification image, showing that more than 95% of the samples are hollow spheres. In the image, some dumbbells and aggregated spheres were also found. Because, the amorphous WO_3 was coated on the surface of some dumbbell-like or aggregated carbon spheres, resulting in the replica of the templates' contour. Thus, hollow dumbbells and aggregated spheres were formed after the templates were removed. The enlarged SEM pattern of WO_3 spheres is shown in Figure 7.77b, in which all

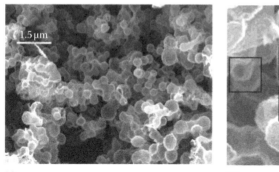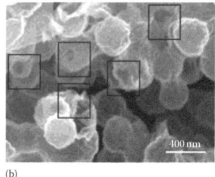

(a)   (b)

FIGURE 7.77
SEM images showing the spherical morphology of WO₃. (a) Low-magnification images. (b) High-magnification images. A small opening in the shell wall of some spheres has been indicated clearly in part (b). (Reprinted with permission from Li, X.-L., Lou, T.-J., Sun, X.-M., and Li, Y.-D., Highly sensitive WO₃ hollow-sphere gas sensors, *Inorg. Chem.*, 43, 5442–5449. Copyright 2004 American Chemical Society.)

of the spheres had uniform morphology and almost the same dimension of about 400 nm. As marked in the image, most spheres had a partial cave-in or a small opening in their shell wall. The composition of the WO₃ hollow spheres was evaluated by standard EDX spectrum. It was found that these hollow spheres were composed of tungsten and oxygen, and the quantitative analysis results indicated that the molar ratio of tungsten and oxygen was about 1:2.98, coincident with XRD and other analyses. It is further reported that these hollow spheres showed good sensitivity to alcohol, acetone, carbon disulfide, and other VOC of specific nature.

Gu et al. [156] reported WO$_x$ nanowires on tungsten metal tips. These tips were prepared by electrochemical etching of tungsten wires by using 1 M KOH solution. SEM studies of the tip morphology showed many nanowires protruding from the tip as shown in Figure 7.78. Flat plates of tungsten, reduced with hydrogen gas gave fine nanowires of tungsten oxide as shown in Figure 7.79a and b. Here the electrochemical etching was carried out for longer periods varying from 1 to 3 h. The prepared tungsten oxide nanotubes or nanowires are expected to have a greater potential for photochromic or electrochromic applications in addition to their application as gas sensors.

Quasi-aligned single-crystalline W₁₈O₄₉ nanotubes and nanowires were synthesized by Li et al. [137], under different atmospheric pressures. For this they have developed a simple technique, to grow them, by using tungsten filaments in air and adjacent *in situ* evaporation. Silicon and tantalum substrates were used to grow these nanostructures. The diameters of these nanotubes were between 150 and 250 nm, while those of the nanowires were found to be in the range of 20–100 nm. Here, a deficiency in vapor supply plays an essential role in the nucleation of these tubular nanostructures. As-grown nanotubes have uniform lengths, normally less than 3 μm, and

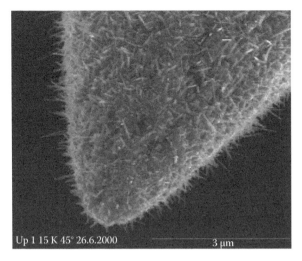

FIGURE 7.78
SEM image of WO$_x$ nanowires at a tungsten tip that was prepared by electrochemical etching. (Reprinted with permission from Gu, G., Zheng, B., Han, W.Q., Roth, S., and Liu, J., Tungsten oxide nanowires on tungsten substrates, *Nano Lett.*, 2, 849–851. Copyright 2002, American Chemical Society.)

(a)   (b)

FIGURE 7.79
SEM images of tungsten oxide nanowires without coating grown from tungsten plates by electrochemical etching for (a) 1 h and (b) for 3 h. (Reprinted with permission from Gu, G., Zheng, B., Han, W.Q., Roth, S., and Liu, J., Tungsten oxide nanowires on tungsten substrates, *Nano Lett.*, 2, 849–851. Copyright 2002, American Chemical Society.)

are quasi-aligned on the tantalum substrates. The top-view SEM image, in Figure 7.80, shows that many tubular nanostructures are grown on the substrate, while some thinner nanowires are also present. This is depicted in Figure 7.80a in a tilted view of the as-grown nanotubes and Figure 7.80b shows the top view. The nonuniformity in the diameters is clearly visible

(a)   (b)

FIGURE 7.80
SEM images of the quasi-aligned $W_{18}O_{49}$ nanotubes: (a) tilted view and (b) top view. (From Li, Y., Bando, Y., and Golberg, D.: Quasi-aligned single-crystalline $W_{18}O_{49}$ nanotubes and nanowires. *Adv. Mater.* 2003. 15. 1294–1296. Copyright Wiley-VCH Verlag GmbH & Co. KGaA. Reproduced with permission.)

from this picture. As discussed, deficiency in vapor supply plays an essential role in the nucleation of these tubular nanostructures. XRD results and TEM observations revealed that these nanowires are single-crystalline monoclinic $W_{18}O_{49}$ structures.

Slightly different structures but with different geometrical shapes are reported for these $W_{18}O_{49}$ compositions. Pine-tree texture of $W_{18}O_{49}$ crystalline and hollow fibers are reported, by Hu et al. [138], by heating WS_2 precursor in the presence of oxygen. This resembles a thick bushy growth of a plant with many needle type structures. SEM study showed that the needles consist of long straight hollow fibers, in the range of 3–10 µm of outer diameter, sub-micron thick wall, and a whopping length of the order of 0.1–2 mm. The whole arrangement gives an impression of a tree-like structure as shown in Figure 7.81a. Figure 7.81b shows an enlarged view of $W_{18}O_{49}$ hollow fiber. The perfect surface smoothness of this fiber and the lengths at which they grow may find a suitable application in the field of gas sensors. Figure 7.81c shows the EDX analyses of individual fibers. The recording showed the presence of only the tungsten and oxygen without any traces of sulfur from the used precursor. Copper signal is initiated from TEM grid. XRD measurements showed that these fibers are in a monoclinic phase, corresponding stoichiometrically to $W_{18}O_{49}$ or more often referred as $WO_{2.72}$.

Figure 7.82a shows the effect of working temperature, in the range of 40°C–250°C, on the sensor response of the tungsten oxide nanostructures toward 1000 ppm H_2S gaseous species. The $WO_{2.72}$ nanowires show the highest values of response toward H_2S while the WO_3 nanoparticles show the least response at all the temperatures, studied. All the nanostructures, however, show a response of 150 at 50°C to 1000 ppm of H_2S, but were found to be reasonably good response value even at temperatures between 50°C and 100°C [135]. The concentration-variation of response of the tungsten oxide

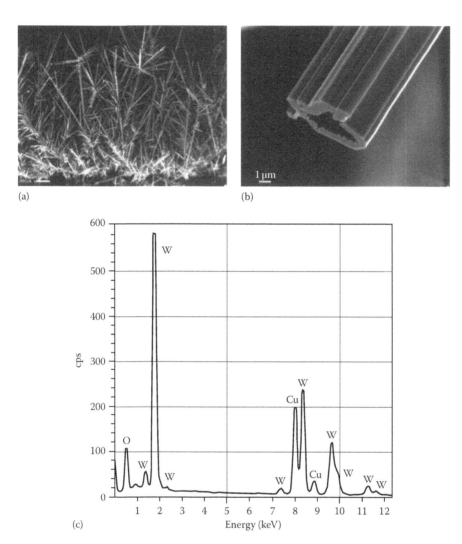

FIGURE 7.81
(a) The pine-tree texture of $W_{18}O_{49}$ crystalline fibers. (b) SEM image of $W_{18}O_{49}$ hollow fiber. (c) EDX analysis of individual fibers. W and O profiles are present as indicated. Copper signal is initiated from TEM grid. (With kind permission from Springer Science+Business Media: *Appl. Phys. A: Mater. Sci. Process.*, Generation of hollow crystalline tungsten oxide fibres, 70, 2000, 231–233, Hu, W.B., Zhu, Y.Q., Hsu, W.K., Chang, B.H., Terrones, M., Grobert, N., Terrones, H., Hare, J.P., Kroto, H.W., and Walton, D.R.M.)

nanostructures at 250°C is shown in Figure 7.82b. In the concentration range of 50–100 ppm, the response is generally satisfactory. The recorded values of response are 392, 121, and 50 to 50, 10, and 1 ppm of H_2S at 250°C in the case of the $WO_{2.72}$ nanowires. The response of 121 of the nanowires to 10 ppm of H_2S is significant since the characteristic bad odor of H_2S gas manifests

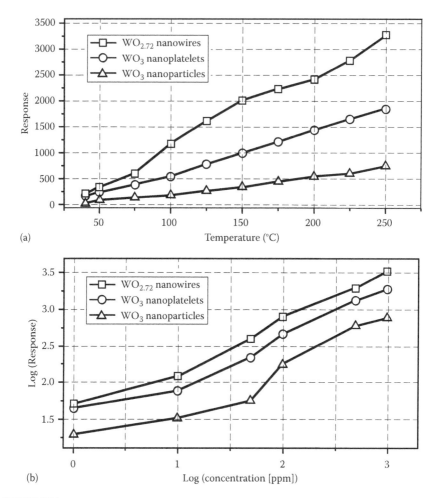

FIGURE 7.82
A comparison of the response values of tungsten oxide nanostructures with (a) temperature (to 1000 ppm H_2S gas) and (b) H_2S gas concentration at 250°C. Measurements were carried out using test gas mixed with dry air. (Reprinted from *Sens. Actuators*, B128, Rout, C.S., Hegde, M., and Rao, C.N.R., H_2S sensors based on tungsten oxide nanostructures, 488–493, Copyright 2008, with permission from Elsevier B.V.)

above this concentration. The response and recovery time curves of these oxide nanoparticles, nanoplatelets and nanowires between 40°C and 250°C is shown Figure 7.83a and b. The response times vary in the 55–100 s range for the nanoplatelets and nanowires, whereas for the nanoparticles the response time is between 80 and 130 s. Thus, the nanoparticles show slower response compared to the nanowires and platelets. The recovery times of all the nanostructures are in the range of 18–40 s depending on the temperature maintained.

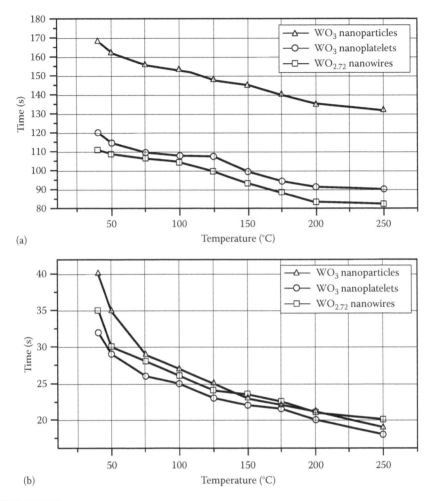

(a)

(b)

FIGURE 7.83
Temperature variation of (a) response and (b) recovery times (to 1000 ppm H_2S) of tungsten oxide nanoparticles, nanoplatelets, and nanowires. (Reprinted from *Sens. Actuators*, B128, Rout, C.S., Hegde, M., and Rao, C.N.R., H_2S sensors based on tungsten oxide nanostructures, 488–493, Copyright 2008, with permission from Elsevier B.V.)

Using nanoparticle thin films for sensing applications is a promising method to improve the response time and the sensitivity of the sensors [256]. The sensing mechanism of nanoparticles with respect to CO and NO measurements is attributed to interactions between different species on the particle surface. It has been established that both gases react with pre-adsorbed oxygen or oxygen from the lattice sites, and release free electrons into the conduction band thus bringing an appreciable change in the electrical conductivity of the sensing material. This change in the conductivity is measured as a signal.

The reducing chemical surface reactions of NO and CO with the WO_3 particles are displayed in the following equations:

$$W_W^X + O_O^X + NO \Rightarrow W_W'' + V_O^{\cdot\cdot} + NO_2 \tag{7.1}$$

$$W_W^X + O_O^X + CO \Rightarrow W_W'' + V_O^{\cdot\cdot} + CO_2 \tag{7.2}$$

Additionally, the disproportionation of NO at the particle surface takes place:

$$NO \Rightarrow N_2O + NO_2 \tag{7.3}$$

Therefore, further oxidizing reactions occur:

$$W_W'' + V_O^{\cdot\cdot} + N_2O \Rightarrow W_W^X + O_O^X + N_2 \tag{7.4}$$

It has been reported that tungsten oxide can crystallize into several polymorphs when exposed to high temperatures.

Lee et al. [256] reported solution-based preparation of soluble and highly crystalline tungsten oxide nanorods of varying lengths. This is a simple large-scale preparation of soluble and highly crystalline nanorods of varying sizes by a mild, solution-based colloidal approach. The length of these nanorods can be easily varied by simple changes in the reaction parameters. Shorter nanorods of 25 ± 6 nm in length (with an aspect ratio of ≈ 10) were obtained at the reaction temperature of 250°C. Longer nanorods of 130 ± 30 nm in length (with an aspect ratio of ≈ 20) were prepared at 270°C by using 12 equiv of oleylamine instead of 16 equiv. At reaction temperatures below 250°C, no nanorod formation was observed, and contamination by platelets was observed at reaction temperatures above 270°C. Longer reaction time caused little effects on the lengths of these nanorods.

Table A.18 explains the details of the tungsten oxide nanostructures and the process of synthesis. The material exhibits gas-sensing properties both in its stoichiometric and non-stoichiometric forms. Different composition levels of tungsten and oxygen are listed in this table, such as $WO_{2.72}$, WO_x ($2.6 \geq x \leq 2.8$), WO_3, WO_{3-x}, WO_3 (In), WO_3 (Pd), and W-Ti-O are listed along with the gaseous species that they respond. The gaseous species are CH_3COCH_3, NH_3, C_6H_6, CS_2, CO, C_2H_5OH, HCHO, H_2, H_2S, CH_3OH, CH_3CN, nitrogen oxides, O_3, petrol (gasoline), petroleum ether, and SO_2. Some of these sensors are very quick to respond particularly the oxides doped with indium metal [158]. This material also needs slightly higher working temperature for their operation and such temperatures are expected to realize in silicon technology based MEMS structures. These forms of nanostructures and their detection capabilities have drawn the attention of many researchers worldwide and the sensors developed on these oxide materials are expected to play an important in near future.

7.18 Vanadium Oxides

Vanadium forms numerous and frequently complicated compounds because of its variable valencies. It exhibits four oxidation states: +2, +3, +4, and +5. It is amphoteric, mostly basic in the lower oxidation states, acidic in the higher states. Vanadium does not tarnish in air readily but when heated it combines with oxygen, nitrogen, and sulfur. The oxides corresponding to the four oxidation states are VO, V_2O_3, VO_2, and V_2O_5.

Vanadium oxides constitute a fascinating class of materials with outstanding physical and chemical properties. They are used in many technological applications, such as in electrical and optical switching devices, light detectors, critical temperature sensors, write-erase media, and in heterogeneous catalysis [257]. In the field of nanoscience, vanadium oxide nanolayers have interesting potential as model oxide systems with a high flexibility in structural and electronic behavior.

Helium (He) gas as one of the noble gases is unreactive, colorless, and odorless. Due to such chemically stable properties, it is very difficult to detect the gas by using the well-known chemical reactions or modulations of electronic states by chemical doping. Yu et al. [160] reported, for the first time, a helium gas sensor based on the vanadium pentoxide (V_2O_5) nanowire featuring conductance variations with various pressures of the helium gas. These V_2O_5 nanowires were synthesized by using sol–gel technique of polycondensation of vanadic acid in water. By repeated injection and evacuation technique, measurements were carried out in nanowires paving a way for the development of nanoelectronic devices. The physical adsorption is a probable mechanism to understand the reaction of helium gas to the nanowires. Because the nanowires and nanotubes have a large surface area, the physical adsorption can happen easily in the nanowires when compared to the normal film or bulk samples. All these measurements were carried out at room temperature by using Si–SiO$_2$ as the substrate for measurements. The response time of 10 s is an important parameter for this V_2O_5-based sensor. This may open new avenues for helium gas detection, particularly in the vacuum techniques. Details of these sensors are shown in Table A.19. The range of detection limits is shown as 0.1–100 mbar.

7.19 Zinc Oxides

Zinc is a fairly active metal chemically. In chemical compounds zinc exhibits almost exclusively a +2 or divalent oxidation state. A few zinc(I) compounds have been reported, but never any compounds of zinc(III) or higher. This metal also forms many coordination compounds.

Zinc oxide, particularly ZnO phase, is a multifunctional semiconducting metal oxide and one of the most promising materials for gas-sensing application, next to tin oxides. ZnO belongs to the II–VI group and wurtzite ZnO has a hexagonal structure, with space group P63mc, which is one of the important semiconductors, mainly due to its wide direct bandgap (3.37 eV) and large excitation binding energy (60 meV). Their use as a gas sensor, in which the surface conductivity changes, in response to adsorbed gases, made them an ideal candidate in the early days of surface science. ZnO arrays or films have been an active research field as early as the 1960s because of their applications as sensors, transducers, and catalysts. Point defects on ZnO surfaces are extremely important, for gas-sensing applications, as they produce very large changes in their surface conductivity. These changes occur at the surface of the grains, as a result of charge transfer, and band bending caused by the adsorbates. The dominant defects identified in these films are mainly the oxygen vacancies. Heating the films to high temperatures generally creates these vacancies. Other technological applications of ZnO include optoelectronic devices such as nanolasers and field emission devices (FED) as well as UV absorber and photoanodes for photovoltaics and DSSCs. ZnO is also a promising material for photonic and optical devices such as photodetector [258]. Recognizing the importance of this material efforts are being made for growing large size wafers, like silicon, using hydrothermal method, melt growth, and seeded vapor transport growth techniques [259].

ZnO is sensitive to many gases, and has satisfactory thermal stability during the usage. Its gas selectivity can be improved by doping additives and catalysts. But, its working temperature is rather high, normally 400°C–500°C, and its gas selectivity is rather poor [260]. In recent years, the studies on various nanostructure ZnO materials have increased, mainly focusing to improve its preparation techniques and to reduce its operating temperatures to the levels of 300°C.

ZnO is naturally an *n*-type semiconductor mainly due to the presence of intrinsic defects, such as oxygen vacancies and zinc interstitials. They form shallow donor levels with ionization energy about 30–60 meV. It is also reported that the *n*-type conductivity is due to hydrogen impurity introduced during the growth process. Till date, various types of dopants, such as group-III (Al, Ga, In), group-IV (Sn), group-V (N, P, As, Sb), group-VI (S), and transition metal (Co, Fe, Ni, Mn) have been implanted into ZnO nanostructures [18]. Doping group-III and group-IV elements into ZnO has proved to enhance its *n*-type conductivity. On the other hand, *p*-type ZnO has been investigated by incorporating group-V elements. In addition, co-doping nitrogen with group-III elements was found to enhance the incorporation of nitrogen acceptors in *p*-ZnO by forming N–III–N complex in ZnO. *n*-type ZnO is easily realized via substituting group-III and group-IV elements or by incorporating excess zinc. By using vapor trapping configuration, it was shown that the electrical properties of ZnO nanowires can be tuned by adjusting synthesis conditions to generate native defects of oxygen vacancy and zinc interstitials as desired.

In the past decade, various chemical and physical deposition techniques have been employed to create oriented arrays. For instance, CVD, physical vapor deposition (PVD), and pulsed laser deposition (PLD) are a few to mention. These methods, however, often suffer from the disadvantage of introducing metal catalysts and requiring high temperature, which could make the synthesis procedures more complex and introduce catalyst impurities to influence the properties of these ZnO nanoarrays.

Zinc oxide nanostructures are synthesized by different approaches varying from simple chemical routes and by sophisticated techniques. Table A.2 lists the details of the shape of nanostructures and the techniques used to synthesize them. The techniques are catalysis-driven MBE technique [186], catalyst free solution route [184], cathodic vacuum arc deposition [55], condensation from vapor phase [87], controlled solid–vapor process [60], co-sputtering using ZnO and copper targets in argon environment [190], deposition by MBE using zinc and O_3/O_2 plasma discharge method [185], direct immersion of zinc foil into NaCl solution (with CH_3COOH) [73], direct immersion of zinc foil with formamide-induced growth [73], domestic microwave oven using zinc and steel-wool at 2.45 GHz [182], electrochemical and solution deposition technique [175], electrochemical route directly on zinc substrate [163], evaporation of ZnO mixed with graphite powder in argon environment [170], fabricated within nanochannels of porous anodic alumina templates by template wetting process [189], hydrothermal method [188], low temperature thermal oxidation of zinc surfaces directly in air [177], modified CVD syntheses approach with vapor trapping method [173], MBE using metal with O_3/O_2 plasma discharge sources and gold droplets as catalyst [165], novel low temperature aqueous chemistry technique [72], one-step process by direct heating of zinc powder without any catalyst [168], physical evaporation method and vertically aligned on templates [179], RF sputtering technique by using pressed powder target [164], selective vapor–solid growth using a seed layer [166], simple thermal evaporation without vacuum in argon environment [178], simple wet chemical route at room temperature [167], solution route employing dodecyl benzene sulfonic acid sodium salt as a modifying agent [187], synthesized through wet chemical route [169], temperature ramping process by using zinc metal in the presence of oxygen and argon ambient [171], thermal chemical reactions and vapor transport deposition method in air [183], thermal evaporation and oxidation of metallic zinc powder without metal catalyst or additives [172], thermal evaporation of zinc powders with oxygen carrier gas ambient [174], thermal evaporation of zinc under controlled condition without a metal catalyst [34], two-step oxygen injection method without catalysts [176]. Synthesized nanostructures include flower-like nanosheet arrays [73], flower-like nanostructures [169], hexagonal nanorods [73], nano-belts [60,163,164], nanocantilevers [60], nanocrystalline thin films [190], nanocrystals [60], nanofibers [182–184], nanorods [60,163,165–168,185–188],

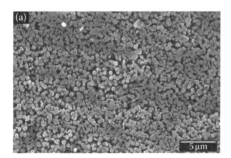 

FIGURE 7.84

SEM images of ZnO nanorod arrays synthesized by direct immersion of zinc foil into NaCl solution at 120°C for 16 h: (a) low-magnification view and (b) high-magnification view. (Reprinted from *J. Cryst. Growth*, 310, Yan, C. and Xue, D., Solution growth of nano- to microscopic ZnO on Zn, 1836–1840, Copyright 2008, with permission from Elsevier B.V.)

nanotubes [189], nanowall networks with a honeycomb-like pattern [170], nanowires [34,87,171–180], ultrathin films [55], and a wide variety of other nanoshapes [72].

Yan and Xue [73] have adopted a strategy to synthesize ZnO nanorod arrays on a zinc surface by employing NaCl solution corrosion-based approach. The selection of zinc foil as the substrate for the growth of well-orientated ZnO arrays was used because of lattice matching between ZnO and zinc crystals. This approach allows growing the well-aligned ZnO nanorod arrays. Since zinc foil is an electrically conductive material and is highly useful to utilize the aligned ZnO nanorods for electronic and optoelectronic device applications. Figure 7.84a shows the general morphology of the ZnO nanorod arrays that are densely packed on the zinc foil substrate. These SEM images of the ZnO nanorod arrays synthesized by direct immersion of zinc foil into NaCl solution at 120°C for 16 h. It is clearly seen from Figure 7.84b that well-aligned nanorod arrays, at high-magnification view, are grown along the [001] crystal direction in a perpendicular fashion onto the substrates. TEM analysis of these nanorods indicated that it is straight with a diameter of 500 nm grown up to a length of 1–3 mm. HRTEM image of the individual ZnO nanorod revealed that only the fringes of the (002) plane with a lattice spacing of about 0.26 nm were observed, confirming that [001] is the growth direction of these nanorods.

The size and shape of ZnO nanorod arrays can be effectively tuned by adjusting the concentration of NaCl solution and the pH value of reaction solution. As shown in Figure 7.85a and b, hexagonal ZnO rods with a diameter of 2 µm can be obtained when the concentration of NaCl solution is adjusted to a value of 0.45 mol/L. This approach confirms the versatility of the syntheses of these ZnO nanostructures. It is also possible to create novel 2-D pattern of flower-like ZnO nanosheets by formamide-induced sequential nucleation and growth on zinc foil in the $Zn(OH)_4^{2-}$ solution without NaCl. It has been demonstrated that secondary growth on initially formed nuclei was needed for the growth of complex patterns of ZnO nanocrystals as shown in the figure.

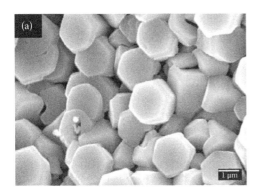
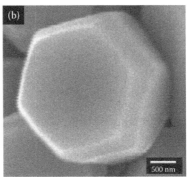

FIGURE 7.85
(a) SEM image of hexagonal ZnO rods and (b) at a higher magnification. (Reprinted from *J. Cryst. Growth*, 310, Yan, C. and Xue, D., Solution growth of nano- to microscopic ZnO on Zn, 1836–1840, Copyright 2008, with permission from Elsevier B.V.)

Pd-coated ZnO nanorods appear well suited to detection of ppm level concentrations of hydrogen at room temperature. The resulting structures show a change in room temperature resistance upon exposure to hydrogen concentrations in nitrogen atmosphere, in the range of 10–500 ppm, approximately by a factor of 5 larger than without the palladium coating. Wang et al. [185] have demonstrated the advantages of using palladium surface promoter for these nanorods. ZnO nanorods were deposited by MBE technique using pure zinc metal and an O_3/O_2 plasma discharge as the source materials. Figure 7.86a shows the SEM of multiple nanorods deposited by MBE technique. Figure 7.86b shows the photograph of the nanorods contacted with Al/Pt/Au

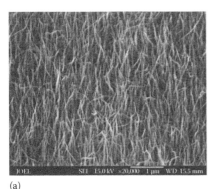

(a)   (b)

FIGURE 7.86
(a) Scanning electron microphotograph of ZnO multiple nanorods deposited by MBE technique. (b) Photograph of the nanorods contacted by Al/Pt/Au electrodes. The ZnO chip has an edge length of about 5 mm in this figure. (Reprinted with permission from Wang, H.T., Kang, B.S., Ren, F., Tien, L.C., Sadik, P.W., Norton, D.P., Pearton, S.J., and Lin, J., Hydrogen-selective sensing at room temperature with ZnO nanorods, *Appl. Phys. Lett.*, 86, 243503. Copyright 2005, American Institute of Physics.)

electrodes. The addition of palladium metal appears to be effective in catalytic dissociation of the H_2 molecule to atomic hydrogen. Though the response is relatively slow, ranging to several minutes, the recovery time of these sensors were found to be very fast. The measured value was found to be less than 20s in many cases. The exact mechanism is not very clear. This rapid and easy recoverability of these sensors make the Pd-coated nanorods suitable for practical applications in hydrogen-selective sensing at ppm levels operating at room temperature where they consume negligible power. This is a novel approach to detect the concentration levels of hydrogen at ppm levels in general ambient.

Lv et al. [187] have grown well-crystalline ZnO nanorods by a simple solution route employing dodecylbenzene sulfonic acid sodium salt as a modifying agent. The as-grown ZnO nanorods were about 95nm in diameter and a few micrometers long. Sensors based on these nanorods were developed by using a paste rich in these nanotubes coated onto a tubular Al_2O_3 structures. The sensors exhibited high responses and good selectivity to ethanol and benzene vapors. The response to 0.01 ppm benzene and 1 ppm ethanol was reported by them. The response times and the recovery times for ethanol gas were no more than 10s when operating at 450°C. These results demonstrate that ZnO nanorods have a promising application in highly sensitive benzene and ethanol gas sensors which are categorized under VOC group.

Khan et al. [168] have reported a simple and effective way to fabricate well-aligned ZnO nanorods using the catalyst-free thermal evaporation method. They exhibited uniform size and in distribution, as shown in Figure 7.87a and b. A small portion is enlarged for a better view of these nanorods in Figure 7.87b inset. Single crystal silicon substrates were used to grow these nanorods. Distribution of these nanorods is found to be very uniform. Analysis of these nanorods, using XRD spectra of as grown nanorods, indicated that they are

(a)   (b)

FIGURE 7.87
SEM image of aligned nanorods grown on Si (100) substrate at temperature 700°C for 30 min: (a) side view and (b) top view of the same sample. The scale bar of the inset in part (b) is 200 nm. (Reprinted from *Phys. E Low Dimens. Syst. Nanostruct.*, 39, Khan, A., Jadwisienczak, W.M., Lozykowski, H.J., and Kordesch, M.E., Catalyst-free synthesis and luminescence of aligned ZnO nanorods, 258–261, Copyright 2007, with permission from Elsevier B.V.)

of high quality single crystalline structures. The largest diffraction peak was centered at $\theta = 34.45°$ and this corresponds to ZnO (002) wurtzite phase and shows a good alignment of nanorods along [0001] direction. The growth of the large-scale aligned ZnO nanorods at relatively low temperature and that too without any catalyst is an attractive alternative to the metal-catalyst-assisted VLS method. Getting this type of alignment on silicon substrates is also attractive and this may simplify the compatibility problems for the development of small size integrated gas sensors with advanced silicon fabrication techniques.

Surface-depletion controlled gas-sensing ZnO nanorods are reported by Li et al. [167] and they used simple wet chemical route to synthesize them at room temperature. Figure 7.88a shows a typical SEM image of as-grown ZnO nanorods showing a bush-like assembly. The average diameter is about 15 nm and measuring 1.0 μm long. Figure 7.88b shows the HRTEM image of the ZnO nanorods. These nanorods were analyzed to be single crystalline in nature. Figure 7.88c shows the electron diffraction pattern of these rods and Figure 7.88d is the XRD pattern of these nanorods to confirm their crystalline nature. These sensors exhibited much higher sensitivity and a quick response and recovery times when compared with other sensors. This was attributed due to an almost complete depletion of the nanorods, which is close to two times the Debye length of ZnO. These types of structures offer very large surface areas, which is essential for gas-sensing applications.

Yang et al. [163] have developed an electrochemical route for the synthesis of highly ordered ZnO ultrathin nanorod and hierarchical nanobelt arrays on zinc substrate. This technique offers better control, mild electrochemical conditions, and ultra fine features and is suitable for integrated device structures. Figure 7.89a shows the SEM images of ZnO nanobelt arrays. An enlarged section of the figure, Figure 7.89b, shows cone-shaped bundles of ultrathin ZnO nanobelts with an average length of about 400 nm uniformly distributed over a wide area and are well aligned vertically on the zinc substrate, resulting in a hierarchically ordered array. The inset in this figure shows the enlarged image of the nanobelt array. Each cone-shaped nanobundle comprises tens of neighboring nanobelts with their top ends lumped together as seen in Figure 7.89c. Figure 7.89d through f shows the TEM images of the ZnO nanobelts and FFT pattern of the nanobelt. These nanostructure arrays possess a large surface-to-volume ratio and are in a unique configuration making them attractive candidates for gas and chemical sensing applications. Hydrogen gas sensor based on these ZnO nanobelt array exhibited excellent sensitivity, rapid response, and good reproducibility even at room temperature operation.

Nanoarrays of ZnO on zinc foil were also reported by Yan and Xue [73]. Figure 7.90a shows arrays of flower-like ZnO nanosheets formed by formamide-induced growth in the $Zn(OH)_4^{2-}$ solution. From the high-magnification image, as shown in Figure 7.90b, it is clearly seen that the flower-like crystal is composed of many 2-D nanosheets that are simultaneously grown from

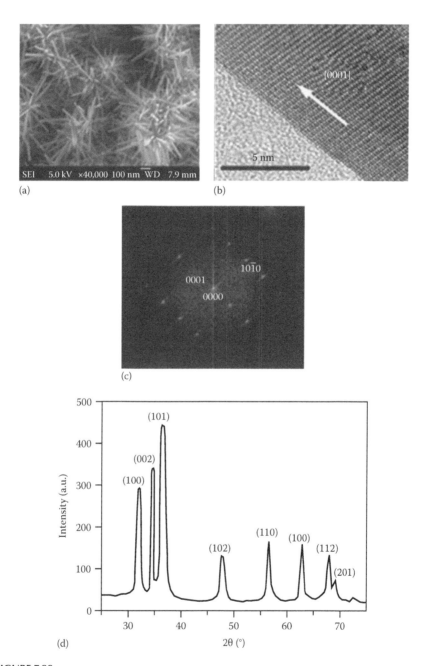

FIGURE 7.88

(a) SEM image of ZnO nanorods synthesized by a simple wet chemical route. (b) HRTEM image of the ZnO nanorods. (c) Electron diffraction pattern of the rods. (d) XRD pattern of the ZnO nanorods. (Reprinted with permission from Li, C.C., Du, Z.F., Li, L.M., Yu, H.C., Wan, Q., and Wang, T.H., Surface-depletion controlled gas sensing of ZnO nanorods grown at room temperature, *Appl. Phys. Lett.*, 91, 032101. Copyright 2007, American Institute of Physics.)

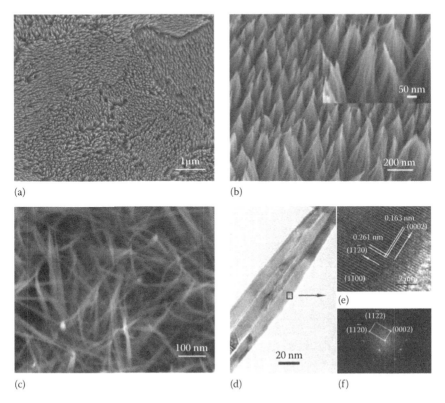

FIGURE 7.89

(a through c) SEM images of a ZnO nanobelt array grown on zinc substrate. (d and e) TEM images of the ZnO nanobelts. (f) FFT pattern of the nanobelt in (e). Inset of (b) is an enlarged image of the ZnO nanobelt array. (Reprinted with permission from Yang, J., Liu, G., Lu, J., Qiu, Y., and Yang, S., Electrochemical route to the synthesis of ultrathin ZnO nanorod/nanobelt arrays on zinc substrate, *Appl. Phys. Lett.*, 90, 103109. Copyright 2007, American Institute of Physics.)

pre-agglomerated seeds. This is due to the contribution of formamide molecules present in the reaction system. Without the formamide molecules, only 1-D ZnO nanorod arrays are observed. When formamide molecules are introduced into the initial reaction system, they can accelerate the oxidation process on the zinc surface due to the formation of a zinc-formamide complex. This complex transforms into ZnO at elevated temperature through thermal decomposition. Due to the enhanced oxidation of metal zinc, numerous nuclei are quickly formed and secondary growth produces flower-like ZnO nanosheets from those already formed nuclei in $Zn(OH)_4^{2-}$ solution. Appropriate volume of formamide is a critical factor, in this case, to produce flower-like ZnO nanosheets. When the volume of formamide is decreased from 15 to 5 mL, flower-like ZnO nanorods, as shown in Figure 7.90c and d, are obtained instead of ZnO nanosheets. Flower-like ZnO nanosheets and nanorods can be obtained in the presence of 15 and 5 mL formamide in the

FIGURE 7.90
SEM image of ZnO nanoarrays grown on the zinc foil substrate in the $Zn(OH)_4^{2-}$ solution at 120°C with different volumes of formamide: (a) and (b) with 15 mL, (c) and (d) with 5 mL. (Reprinted from *J. Cryst. Growth*, 310, Yan, C. and Xue, D., Solution growth of nano- to microscopic ZnO on Zn, 1836–1840, Copyright 2008, with permission from Elsevier B.V.)

$Zn(OH)_4^{2-}$ solution, respectively, as shown in Figure 7.90. This type of nanostructures is expected to play a key role in gas-sensor application due to their large exposed surface area where surface interactions are possible.

ZnO is a great example of the capability of low-temperature aqueous chemical growth to design, onto many different kinds of substrates, various ordered architectures including wires, rods, tubes and star-shape at nano-, meso-, and microscale. Vayssieres [72] reported the growth of ZnO by heteronucleation onto the substrates and various morphologies and orientation monitoring were obtained by experimental control of the chemical composition of the precipitation medium as shown in Figure 7.91. Growth was carried out at a temperature of 90°C. The figure shows purpose built crystalline ZnO structures directly grown onto various substrates, such as: ITO, single crystal silicon wafers, glass plates, and polypropylene sheets. These varieties of nanostructures of ZnO are very interesting to study further for possible gas-sensing and other technological applications.

As discussed earlier, till date various techniques have been employed for the preparation of ZnO nanoparticles, which included sol–gel, physical

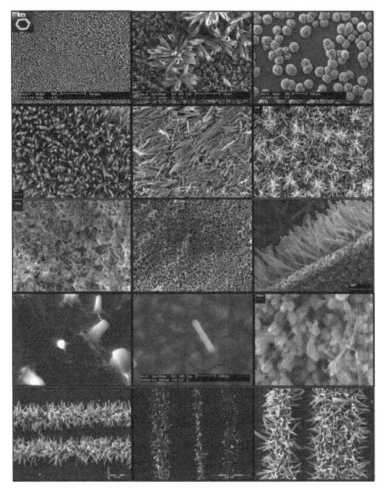

FIGURE 7.91
Purpose-built crystalline ZnO structures grown directly onto various substrates of ITO, single crystal silicon, glass, and polypropylene by controlled aqueous growth technique at a temperature of 90°C. (Reprinted from *Comp. Rend. Chim.*, 9, Vayssieres, L., Advanced semiconductor nanostructures, 691–701, Copyright 2006, with permission from Elsevier SAS.)

evaporation, pulsed laser ablation, and flame pyrolysis. However, the use of suitable organometallic single-source precursors seems highly desired since they can be employed, at the same time, for solid-state and chemical vapor synthesis (CVS) or in a solution even in the nonaqueous medium. Several organometallic precursors were reported, in the open literature, for MOCVD of ZnO thin films and, in fact, these precursors can be used for the preparation of ZnO nanoparticles. Driess et al. [261] came out with a large number of precursors for the formation of ZnO nanoparticles. Dimethyl- and diethylzinc are widely used, for this application, to coat substrates having large

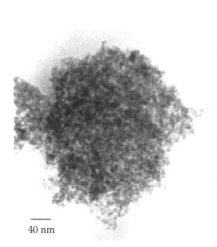
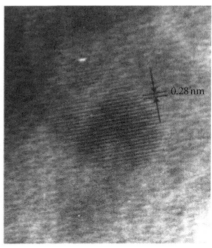

0.28 nm

40 nm

FIGURE 7.92

HRTEM micrographs of ZnO nanoparticles derived from solid-state decomposition of $(MeZnOSiMe_3)_4$. (Reprinted from *Comp. Rend. Chim.*, 6, Driess, M., Merz, K., Schoenen, R., Rabe, S., Kruis, F.E., Roy, A., and Birkner, A., From molecules to metastable solids: Solid-state and chemical vapour syntheses (CVS) of nanocrystalline ZnO and Zn, 273–281, Copyright 2003, with permission from Elsevier SAS.)

surface area and at relatively high growth rates. However, these precursors are highly pyrophoric in nature and one has to use oxygen as co-reactant for the depositions. Extra precautions are necessary to handle this combination during the depositions. The formation of nanocrystalline ZnO particles were found with diameters of 10–12 nm. Further it was concluded from the powder XRD pattern and the very high BET (Brunauer–Emmet–Teller) studies that the surface area of these nanoparticles is in the range of $110 \, m^2/g$. The following Figure 7.92 shows the HRTEM-image of ZnO-nanocrystal clusters with the lattice parameters of wurtzite (d_{hkl} {001} = 2.81 Å). These nanoparticles are derived from solid-state decomposition of precursor volatile zinc solixide $(MeZnOSiMe_3)_4$ [261].

Wan et al. [34] reported the ethanol sensing characteristics of ZnO nanowire fabricated using MEMS technology. These nanowires were synthesized by thermal evaporation of zinc under controlled condition without a metal catalyst. The general morphology of the synthesized ZnO nanowires was studied by using field emission SEM. Figure 7.93a shows these nanowires. It was found that these long nanowires are having a diameter of approximately $25 \pm 5 \, nm$. The synthesized nanowires are composed of oxygen and zinc with an atom ratio of 98%, measured by energy-dispersive x-ray fluorescence (EDX) pattern inferring the existence of some oxygen vacancies, as shown in Figure 7.93b. When the nanowire is exposed to air, an oxygen molecule adsorbs on the surface of the ZnO nanowires and forms an O_2^- ion by capturing an electron from the conductance band. This shows high resistance state

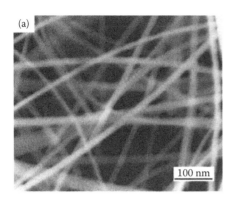

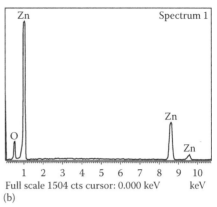

FIGURE 7.93

(a) SEM image of the as-synthesized ZnO nanowires and (b) EDX pattern of the synthesized sample. (Reprinted with permission from Wan, Q., Li, Q.H., Chen, Y.J., Wang, T.H., He, X.L., Li, J.P., and Lin, C.L., Fabrication and ethanol sensing characteristics of ZnO nanowire gas sensors, *Appl. Phys. Lett.*, 84, 3654–3656. Copyright 2004, American Institute of Physics.)

in air ambient. When exposed to a reductive gas it reacts with the surface oxygen species, which decreases the surface concentration of O_2^- ion and increases the electron concentration. The sensor exhibited high sensitivity and fast response and worked at 300°C. It is expected that these ZnO nanowires will be useful to built highly sensitive alcohol sensors.

ZnO nanowire-based CO gas sensors were fabricated, by Hsueh et al. [171], by growing single crystal ZnO nanowires on patterned ZnO:Ga/SiO$_2$/Si templates at various temperatures. It was found that the average length of the nanowires increased while the average diameter of the nanowires decreased as the growth temperature was increased from 600°C to 700°C. It was also found that these nanowires became significantly shorter as the growth temperature was increased. Figure 7.94a and b shows the cross-sectional FE-SEM images of the ZnO nanowires grown at 600°C and 650°C. Similarly Figure 7.94c through e shows the nanowires grown at 700°C, 750°C, and 800°C, respectively. The wires are highly dense and are well-aligned vertically on the substrates. These vertical growths have resulted in an opened up extra surface area. These ZnO nanowires were grown on the conducting ZnO:Ga finger regions while randomly oriented ZnO nanowires were grown on the insulating SiO$_2$ spacer regions. The perfect vertical orientation of these nanowires is very impressive and is evenly distributed in their vertical dimensions. Growing on SiO$_2$ a spacer region, that too at a temperature of 600°C, is a major technological achievement toward the development of micro-gas sensors. This selective growth on SiO$_2$ shows that the devices are compatible with silicon processing technology.

ZnO is a wide bandgap semiconductor that is also suitable for blue optoelectronic applications with ultraviolet lasing action in disordered particles

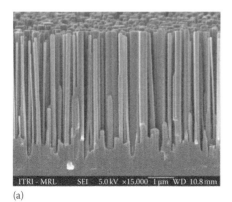
(a)

(b)

(c)

(d)

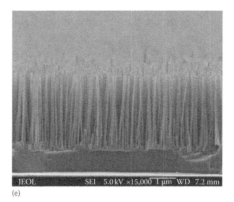
(e)

FIGURE 7.94

Cross-sectional FE-SEM images of the ZnO nanowires grown at (a) 600°C, (b) 650°C, (c) 700°C, (d) 750°C, and (e) 800°C. Magnification values and scale bars are not identical. (Reproduced from Hsueh, T.-J., Chen, Y.-W., Chang, S.-J., Wang, S.-F., Hsu, C.-L., Lin, Y.-R., Lin, T.-S., and Chen, I.-C., ZnO nanowire-based CO sensors prepared at various temperatures, *J. Electrochem. Soc.*, 154, J393–J396, 2007, by permission of ECS—The Electrochemical Society.)

and thin films. Nanowires with flat ends can also be exploited as optical reso-
nance cavities to generate coherent light at nanoscale [68]. Such fine coherent
light sources are expected to find many technological applications. Figure
7.95a shows the schematic illustration of the excitation and detection configuration used for lasing study. Figure 7.95b shows SEM image of vertically
grown nanowires on sapphire substrate. This 2-D array of ZnO nanowires
are grown as uniaxial crystals on the surface of sapphire. Such structures are
highly useful for gas-sensing applications also because of large surface area

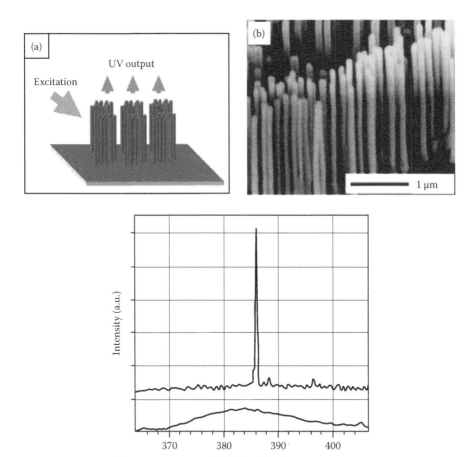

FIGURE 7.95
(a) Schematic illustration of the excitation and detection configuration used for the lasing study.
(b) SEM image of a 2-D array of ZnO nanowires grown as uniaxial crystals on the surface of a
sapphire substrate. (c) The power-dependent emission spectra recorded from a 2-D array of ZnO
nanowires, with excitation energy being below (bottom trace) and above (top trace) the threshold. (From Xia, Y., Yang, P., Sun, Y., Wu, Y., Mayers, B., Gates, B., Yin, Y., Kim, F., and Yan, H.: One-
dimensional nanostructures: Synthesis, characterization, and applications. *Adv. Mater.* 2003. 15.
353–389. Copyright Wiley-VCH Verlag GmbH & Co. KGaA. Reproduced with permission.)

of these nanorods. It is also possible to achieve equally favorable adsorption and desorption behavior at room temperature by illumination, for excitation, the devices with suitable light energy close to the bandgap of this material. Figure 7.95c shows the power-dependent emission spectra recorded from a 2-D array of ZnO nanowires, with the excitation energy being below (bottom trace) and above (top trace) the threshold limits.

The near-band emission, transparent conductivity, and piezoelectricity make 1-D ZnO nanomaterials, one of the most important functional semiconductor oxide nanostructures, serving to field emitters, optoelectronic devices and sensors, etc. Understanding hetero-epitaxial growth of nanometer size ZnO structures such as nanowires, nanorods, and quantum dots on various substrates is important for assembling ordered arrays of these nanostructures into functional ultraviolet optoelectronic and nanoscale electronic devices. One of the promising methods for preserving crystalline order in thin semiconductor films and nanostructures is to grow them epitaxially on lattice-matched substrates. The choice of substrate effects on the final state of strain by directly determining the lattice mismatch and thermal expansion mismatch. Jeon et al. [172] attempted to deposit various ZnO nanowires on different substrates such as single crystal silicon and sapphire by using simple PVD. Figure 7.96 shows the SEM images of the samples grown on three substrates: (1) Si (100) orientation, (2) c-plane (0001) orientation of sapphire, and (3) a-plane (11$\bar{2}$0) of sapphire. After cooling down to room temperature, dark gray–colored deposits were found on the surface of these substrates. All these nanowires on different growth substrates were disorderly arranged and their morphologies were similar. The measured diameter, of these nanowires, was in the range of 30–50 nm and the length was of the order of several microns for all kinds of grown ZnO nanostructures. The nanowires grown on silicon substrate, with (100) crystal orientation, dominantly have (103) growth direction. With c-sapphire substrate, the ZnO nanowires were well crystallized in hexagonal structure without special preference for growth direction. On the other hand, sample grown on a-sapphire substrate, the ZnO (00l) peak is much higher than any other peak, including the substrate peak that indicates that the (002) axes of the ZnO nanowires are well aligned with Al_2O_3 (11$\bar{2}$0) over a large substrate area. The growth density of these nanowires depends on the substrate. In these case nanowires on silicon substrate shows lower value (Figure 7.96a) and followed by sapphire (Figure 7.96b) and alumina (Figure 7.96c). It is not clear whether the surface roughness has any role for this growth density.

Hsueh et al. [176] reported the growth of high-density single crystalline ZnO nanowires on patterned $ZnO:Ga/SiO_2/Si$ templates, palladium adsorbed on nanowire surfaces, and on the fabrication of ZnO nanowire-based ethanol gas sensors. It was reported that the resistivity of the fabricated ethanol sensor decreased while the sensor response increased as they increased the ethanol vapor concentration and operating temperature of the sensing element. Compared with pure ZnO nanowires, it was found that

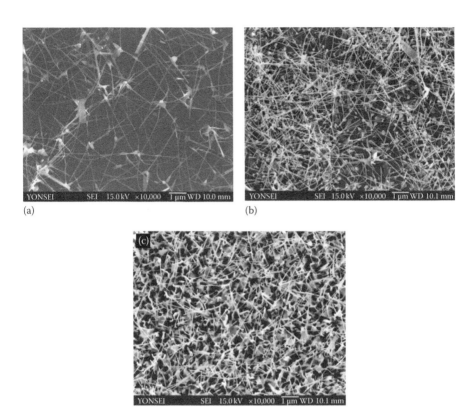

FIGURE 7.96
Top-view SEM images of various ZnO nanowire grown on different substrates: (a) *p*-Si (100), (b) *c*-plane (0001) sapphire, and (c) *a*-plane (11$\bar{2}$0) sapphire substrates. (Reprinted from *Phys. E Low Dimens. Syst. Nanostruct.*, 37, Jeon, K.A., Son, H.J., Kim, C.E., Kim, J.H., and Lee, S.Y., Photoluminescence of ZnO nanowires grown on sapphire (11$\bar{2}$0) substrates, 222–225, Copyright 2007, with permission from Elsevier B.V.)

the ethanol sensor response of the ZnO nanowires with palladium adsorption was much larger. Presence of palladium on surface has enhanced the gaseous interaction. It was highlighted by Hsueh et al. [176] about palladium presence on the ZnO nanowire surface, which enhanced the intrinsic sensing properties of these nanowires. Lu et al. [180] have demonstrated similar nanowires on ZnO:Ga/glass templates. Analysis of these nanowires indicated that they are single crystalline in nature with good crystal quality. The nanowires showed good gas-sensing property at room temperature.

High density, single crystalline ZnO nanowire arrays were successfully prepared by Ren et al. [177] in oxygen or air environment by heating pure zinc metal at temperatures of the order of 350°C and 400°C, which are well below the melting temperature of pure zinc metal. The prepared ZnO nanowires are straight in their shapes and they distributed in high density with slightly random growth directions. The length of these ZnO

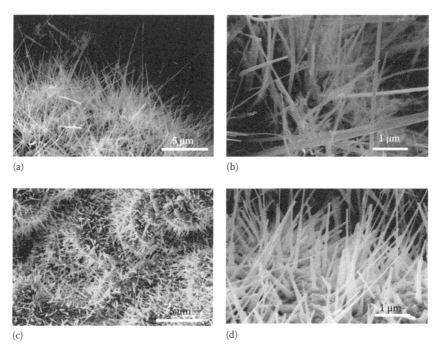

FIGURE 7.97
FE-SEM morphologies of ZnO nanowire arrays. (a) and (b): Grown at 400°C in air for 1 h; (c) and (d): grown at 350°C in flowing oxygen gas for 2 h. (Reprinted from *Mater. Lett.*, 61, Ren, S., Bai, Y.F., Chen, J., Deng, S.Z., Xu, N.S., Wu, Q.B., and Yang, S., Catalyst-free synthesis of ZnO nanowire arrays on zinc substrate by low temperature thermal oxidation, 666–670, Copyright 2007, with permission from Elsevier B.V.)

nanowires varied from several micrometers to over 10 μm, and the diameter of the nanowires ranged from 20 to 150 nm. It is found that these ZnO nanowires formed when oxidized at 400°C and 350°C, but there are no ZnO nanowires observed when oxidation was carried out below 300°C. FE-SEM morphologies of these films, grown at 400°C in air, are shown in Figure 7.97. Figure 7.97a and b shows the FE-SEM morphologies of ZnO grown at 400°C in air for 1 h and Figure 7.97c and d shows the nanowires grown at 350°C in flowing oxygen gas for 2 h. The as-grown ZnO nanowires were found to be single crystalline with a wurtzite structure extending in <110> growth direction. This technique is a promising method for synthesizing high-quality ZnO nanowire arrays and at the same time the technique is highly compatible with the current semiconductor processing technologies. The work also highlights the low-temperature direct oxidation process and also high-quality nanowire arrays for future nanodevices.

Maity et al. [184] have synthesized ZnO nanofibrous thin films by a catalyst free solution route on glass and silicon (400) substrates. XRD study revealed that the formation of these ZnO nanofibers of hexagonal crystalline structure. The texture coefficient of different planes varied with annealing temperature

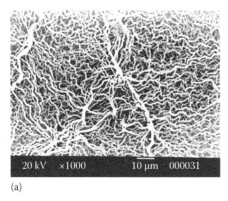
(a)

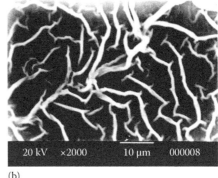
(b)

FIGURE 7.98
SEM micrographs of fiber-structured ZnO thin films deposited on (a) glass and (b) silicon (400) substrates. (Reprinted from *Phys. E Low Dimens. Syst. Nanostruct.*, 25, Maity, R., Das, S., Mitra, M.K., and Chattopadhyay, K.K., Synthesis and characterization of ZnO nano/microfibers thin films by catalyst free solution route, 605–612, Copyright 2005, with permission from Elsevier B.V.)

and that of the (002) plane was the highest for the films annealed at temperature ~873 K. Scanning electron micrograph showed the well formation of ZnO nano/microfibers with an average length of the fibers in the range of 60–80 µm, whereas the average diameter of the fibers is ~500 nm, with an average aspect ratio of around 150. The fiber structure of ZnO film deposited on glass plate is shown in Figure 7.98a and b for the silicon substrate. UV-Vis-NIR spectroscopy measurements showed high transmittance in the visible and near-infrared optical regions. Annealing temperature variation showed changes in the transmittance value and it decreased as the annealing temperature was raised. It is also reported that the bandgap energy decreased as annealing temperature was raised to higher values. The bandgap energy of nanostructured ZnO thin films is reported to be in the range of 3.03–3.61 eV.

Chen et al. [183] have synthesized ZnO crystal fibers on silicon (100) orientation substrate via a simple thermal chemical reaction vapor transport deposition method in air with a mixture of ZnO and carbon powders as reactants. The growth process was carried out at 1100°C in a quartz tube with one side opened to the air. The obtained ZnO consisted of radially grown needle-like nanofibers. These fibers had diameters ranging from 300 nm to 1.5 µm and the maximal lengths up to 1 mm. SEM image of these fibers is shown in Figure 7.99a through c at different magnification values. XRD and Raman spectra studies showed that these fibers were composed of hexagonal wurtzite-phase of ZnO with good crystal quality. The perfect growth of needle-like fibers is expected to play an important role toward the development of integrated gas sensors particularly for open-gate FET structures.

Heo et al. [165] reported a different technique to synthesize ZnO nanorods. Discontinuous gold droplets were used as the catalyst for the growth and they were formed by annealing e-beam evaporated gold thin films (~100 Å)

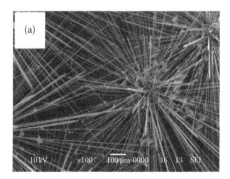
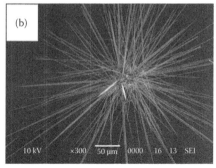
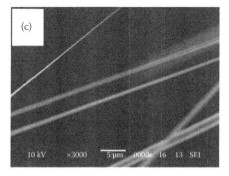

FIGURE 7.99

SEM images of the ZnO crystal fibers on silicon substrate with magnification of (a) 100, (b) 300, and (c) 3000. (Reprinted from *Phys. E Low Dimens. Syst. Nanostruct.*, 21, Chen, B.J., Sun, X.W., Xu, C.X., and Tay, B.K., Growth and characterization of zinc oxide nano/micro-fibers by thermal chemical reactions and vapor transport deposition in air, 103–107, Copyright 2004, with permission from Elsevier B.V.)

on *p*-Si (100) wafers at 700°C. ZnO nanorods were deposited by MBE technique with a base pressure of 5.0×10^{-8} mbar using high purity (99.9999%) zinc metal and an O_3/O_2 plasma discharge as the source chemicals. The zinc pressure was varied between 4.0×10^{-6} and 2.0×10^{-7} mbar, while the beam pressure of the O_3/O_2 mixture was varied between 5.0×10^{-6} and 5.0×10^{-4} mbar. The growth time was ~2h at 400°C–600°C. The typical length of the resultant nanorods was in the range of 2–10mm, with typical diameters in the range of 30–150nm. Selected area diffraction patterns showed that the nanorods are single crystal in nature. They were subsequently released from the substrate by sonication in ethanol and then transferred to SiO_2-coated silicon substrates for further analysis. Electrical conductivity of these nanorods increased by a post-growth anneal in hydrogen ambient and showed a thermally activated current that is insensitive to ambient gases.

In the family of nanobelts, ZnO is probably the most extensively studied structure. Wang [60] has reported the details of ZnO nanobelts and the syntheses of these nanostructures by using controlled solid–vapor process. Thermal evaporation of ZnO powders at 1400°C resulted in ultralong nanobelts. The typical

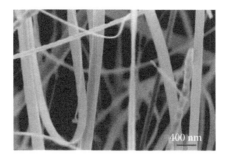

FIGURE 7.100
SEM image of ZnO nanobelts. (Reproduced with permission of Annual Reviews Inc., from Wang, Z.L., Functional oxide nanobelts: Materials, properties and potential applications in nanosystems and biotechnology, *Annu. Rev. Phys. Chem.*, 55, 159–196, 2004, permission through CCC Inc.)

lengths of these ZnO nanobelts are reported to be in the range of several tens to several hundreds of micrometers; some of them even have lengths of the order of millimeters. EDX and XRD measurements have shown that the samples were wurtzite (hexagonal) structure with a lattice constant of $a = 3.249\,\text{Å}$ and $c = 5.206\,\text{Å}$, consistent with the standard values for bulk ZnO. TEM images revealed that the geometrical shape of the ZnO nanobelts is distinct in cross section from the nanotubes or nanowires. Each nanobelt has a uniform width along its entire length, and the typical widths of the nanobelts are in the range of 50–300 nm. HRTEM and electron diffraction studies show that the ZnO nanobelts are structurally uniform and exhibit single crystalline property. SEM images of these nanobelts are shown in Figure 7.100 with uniform structural features.

Tin is an excellent catalyst for growth of ZnO nanostructures. Using tin as a catalyst, Wang et al. [60] have grown aligned ZnO nanowires on a polycrystalline alumina substrate. The SEM image, as shown in Figure 7.101a, clearly displays the reasonable alignment among the nanowires to the substrate. Higher magnification SEM images show that the nanowire has a nonuniform cross section along its length and it becomes sharper toward the tip portions, as seen in Figure 7.101b and c. The very tip has a Sn-rich head and is clearly visible in Figure 7.101b used as a catalyst. Figure 7.101d shows the orientation-ordered epitaxial growth of ZnO nanorods on a large ZnO crystal showing identical orientation.

Seo et al. [189] have adopted a new technique to fabricate ZnO nanotubes. These nanotubes were fabricated within the nanochannels of porous anodic alumina templates by template wetting process. In this method, pore walls of the alumina template were wetted by polymeric ZnO source. After heat treatment procedures, the template was selectively etched off to release the nanotubes. FE-SEM investigations showed that nanotubes have smooth wall morphologies and well-defined diameters, of the order of 200 nm, corresponding to the diameter of the used template, as shown in Figure 7.102. Figure 7.102a shows the top view of ZnO nanotubes partially embedded to the template, Figure 7.102b shows the side wall view of these ZnO nanotubes,

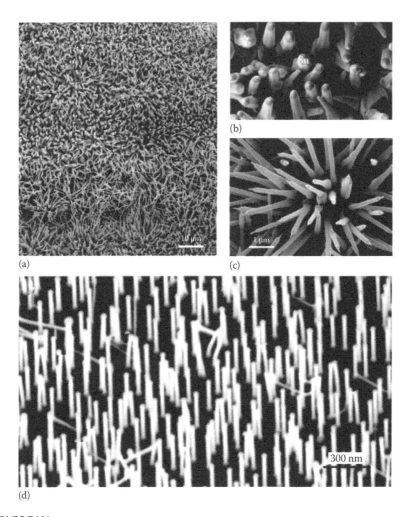

FIGURE 7.101
(a), (b), and (c) Low-magnification and high-magnification SEM images of ZnO nanorods grown on alumina substrate with tin at the growth front. (d) Orientation-ordered epitaxial growth of ZnO nanorods on a large ZnO crystal showing identical orientation. (Reproduced with permission of Annual Reviews Inc. from Wang, Z.L., Functional oxide nanobelts: Materials, properties and potential applications in nanosystems and biotechnology, *Annu. Rev. Phys. Chem.*, 55, 159–196, 2004. With permission through CCC Inc.)

and Figure 7.102c shows the high-magnification FE-SEM image showing open ends of these ZnO nanotubes. XRD measurements showed that these nanotubes have polycrystalline structural properties. These nanotubes fabricated at low temperature are very promising for different applications such as chemical and biological sensors with high sensitivity, optoelectronic devices, and solar cells. This method may be greatly helpful for realizing core-shell nanotube–nanowire heterojunctions.

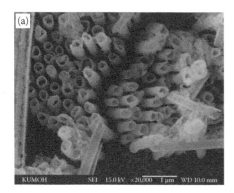

FIGURE 7.102

FE-SEM images of ZnO nanotubes prepared by wetting of nanoporous template: (a) top view of ZnO nanotubes partially embedded to the template, (b) side-wall view of ZnO nanotubes, and (c) high-magnification FE-SEM image showing open ends of ZnO nanotubes. (Reprinted from *Phys. E Low Dimens. Syst. Nanostruct.*, 37, Seo, B.I., Shaislamov, U.A., Ha, M.H., Kim, S.-W., Kim, H.-K., and Yang, B., ZnO nanotubes by template wetting process, 241–244, Copyright 2007, with permission from Elsevier B.V.)

Noble metals in ZnO enhance the catalytic activity of support oxides, keep a particular valence state unchanged, favor formation of active phases, stabilize catalysts against reduction, and/or increase electron exchange rate. The nature of the noble metal grains, namely, their electronic state and distribution, both when applied over the oxide surface, and inside oxide films, is predicted to boost the sensitivity and selectivity of gas detectors by catalyzing specific reactions with the detected gases. Their main advantages include simplicity, compactness, reliability, corrosion resistance, room temperature operation, and stability against perturbations by external electromagnetic fields [262]. The main reason for extending nanosized materials to these applications is the larger active surface area they provide for gas–solid interactions.

Heteroepitaxy of vertically well-aligned ZnO nanowall networks with a honeycomb-like pattern on GaN/c-Al$_2$O$_3$ substrates, by the help of gold catalyst, was reported by Kim et al. [170]. The ZnO nanowall networks with wall thicknesses

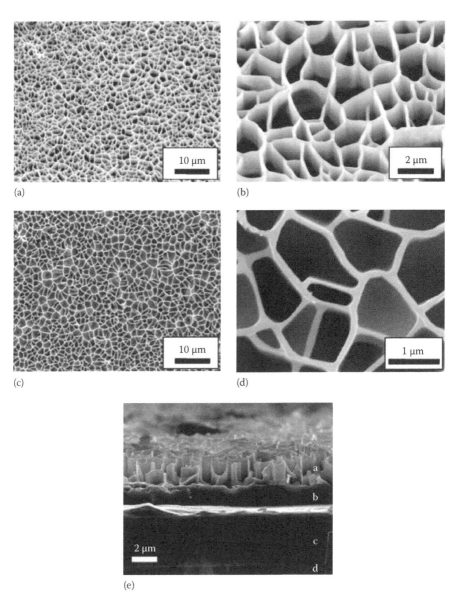

FIGURE 7.103
FE-SEM images of grown ZnO nanowall networks. Tilting-view (30°) FE-SEM images of the
ZnO nanowall networks of (a) low-magnification image and (b) medium-magnification image.
(c, d) Low- and high-magnification plan-view FE-SEM images of the nanowall networks,
respectively. (e) Cross-sectional FE-SEM image of the ZnO nanowall sample. The areas marked
as "a," "b," "c," and "d" are ZnO nanowalls, ZnO thin film, GaN epilayer, and c-Al_2O_3 substrate,
respectively.

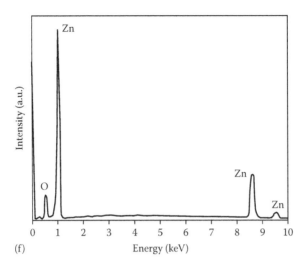

(f)   Energy (keV)

FIGURE 7.103 (continued)
(f) EDS of the area marked "b." (Reprinted with permission from Kim, S.-W., Park, H.-K., Yi, M.-S., Park, N.-M., Park, J.-H., Kim, S.-H., Maeng, S.-L., Choi, C.-J., and Moon, S.-E., Epitaxial growth of ZnO nanowall networks on GaN/sapphire substrates, *Appl. Phys. Lett.*, 90, 033107. Copyright 2007, American Institute of Physics.)

of 80–140 nm and an average height of about 2 μm were grown on a self-formed ZnO thin film during the growth on the GaN/c-Al$_2$O$_3$ substrates. Figure 7.103 shows the ZnO nanowall network. Figure 7.103a shows the FE-SEM images of grown nanowall network with a 30° tilt at low magnification value and Figure 7.103b through d are at higher magnification values. Figure 7.103e shows the cross-sectional view of the nanowall as identified in the figure caption. EDS of the nanowall portion is shown in Figure 7.103f confirms the composition of the nanowall and the presence of zinc and oxygen. It is further reported that the distribution of the nanowall honeycomb-like pattern is nearly uniform over an entire GaN epitaxial layer surface area. The well-aligned ZnO nanowall network is attributed to the good epitaxial lattice match between the ZnO and the hexagonal basal plane of the GaN epilayer on the c-plane Al$_2$O$_3$ substrate. The hydrogen sensing property was reported to be very noticeable since the response time was a few hundred seconds and the resistance change was several tens of percent even in the relatively low temperature of operation at 300°C. The sensing properties of these ZnO nanowall networks suggest that this novel structured material is promising in many application fields, especially as a chemical sensor for detecting noxious gases as well as explosive hydrogen gas.

In the nanocrystals, ZnO is very popular due to the easiness in preparing nanowires and multiple intriguing nanostructures and furthermore to the biocompatibility of this oxide that make it promising for medical and *in vivo* applications. It is well established that the sensing mechanism of ZnO belongs to the surface-controlled type [260]. Its gas sensitivity is relative to grain size, surface state, and active energy of oxygen adsorption and lattice defects. It is normally observed that smaller grain size, specific surface area, and oxygen adsorption quantity enhances the gas sensitivity factor. When the grain size is greater than 40 nm the gas sensitivity decreases quickly because its specific surface area goes down rapidly. The decrease in grain size of ZnO decreases its working temperature due to the increase of surface activity. In general, working temperature of ZnO is 400°C–500°C, but that of nanometer ZnO made by emulsion, it is only around 300°C.

Table A.20 briefs the details of different zinc oxide nanostructures. Electrical resistance variation is a common technique to get the sensor signal. The specific point about this material is its intrinsic behavior. In intrinsic form, it is sensitive to several hydrocarbons and pollutant gases. They include acetone, ammonia, benzene, butane, carbon monoxide, ethyl alcohol, hydrogen, hydrogen sulfide, oxygen, ozone, petrol (gasoline), sulfur hexafluoride, and toluene. Some of the sensors are very quick to respond within few seconds. Others take several minutes. This is another key material for gas-sensing applications. The wide varieties of nanostructures offered by this material are really excellent.

7.20 Zirconium Oxides

Zirconium has good strength at elevated temperatures and it resists corrosion. It will also withstand mechanical damages. This material is predominantly tetravalent in its compounds. Some less stable trivalent compounds, however, are known. Zirconia or zirconium dioxide (ZrO_2) is a hard, white or yellow-brown solid with a high melting point of about 2700°C.

Zirconium oxide based sensors are mainly used as an electrochemical cell configuration and generally they operate at higher temperature ranges. The strategy developed by Chou et al. [54] to fabricate zirconium oxide nanoparticles with hollow structures is remarkable. This nanoshape is unique among all the structures discussed so far. In the preparation of nano to micro ZrO_2 bottles, zirconyl chloride solution and TBA were selected. The zirconyl chloride solution included 10 g zirconyl chloride, $ZrCl_2O \cdot 8H_2O$ and 100 mL water. As shown in Figure 7.104, ZrO_2 particle size can be controlled from spray droplet to reach a nanometer scale and the thickness of the shell particle can be controlled by changing the concentration of the precursor. After the spraying process, removal of TBA and vaporization of water were taken place and then calcinated at 900°C for 4 h. This technique provides a potential way to explore the top-down fabrication of micro- and nano hollow particles from many compounds or colloidal precursor and has promising applications in mass-producing microcapsulate, adsorbent, catalyst, gene gun bullet, and nanostructure materials.

Tan et al. [44] reported the gas-sensing properties of ZrO_2 alloyed with Fe_2O_3. High-energy mechanical ball milling technique was used to synthesize

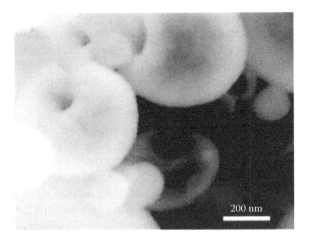

FIGURE 7.104
SEM morphology of micro-zirconium-oxide hollow particle formed by spraying precipitation method with hydrophilic–lipophilic precipitation agents. FE-SEM photography showing nano- to microscale ZrO_2 bottle prepared by using zirconyl chloride solution–TBA system. (Reprinted from *Mater. Sci. Eng.*, A359, Chou, T.-C., Ling, T.-R., Yang, M.-C., and Liu, C.-C., Micro and nano scale metal oxide hollow particles produced by spray precipitation in a liquid–liquid system, 24–30, Copyright 2003, with permission from Elsevier B.V.)

10 nm size grains. Thick films screen printed on ceramic substrates, operating at 120°C–360°C temperatures, showed good response to ethanol vapor. Table A.21 shows the details of this sensing material.

7.21 Mixed Oxides

Mixed nanostructured metal oxides also offer a wide spectrum of gas-sensing materials starting from simple gases like H_2, O_2, humidity, to pollutant gases such as CO, CO_2, and NO_2. The complex nature of the material is the major hurdle for these oxides for formation and in utility. More the number of oxides present in the material more are the difficulties in depositing them, in thin-film form, with good and repeatable stoichiometric ratios. Thick film, pellet, or tubular structure shapes are preferred for these mixed oxides. In thin-film deposition, stoichiometric issues are tough to control.

Perovskites are mixed oxides with the general formula ABO_3, where A is a divalent or monovalent metal and B is a tetra- or pentavalent atom. The interests in perovskite-type oxides are mainly due to the easy modification of their electric properties by the selection of an adequate atom A or B. These nanocrystalline perovskite oxides show very interesting gas-sensing properties. In order to improve the selectivity for a particular application, surface modification is done by proper additives or dopants to the base materials. Cadmium stannate ($CdSnO_3$), a typical compound with a cubic perovskite lattice, shows quite attractive electrical properties [32]. Current years have seen increased interest in gas-sensing devices based on these $CdSnO_3$. It has been observed that $CdSnO_3$ nanopowders synthesized by solid-state reaction and doped with platinum by impregnation technique possesses excellent gas-sensing responses to ethanol.

Xue et al. [263] reported the sensing property of $ZnSnO_3$ nanowires with lengths of several tens micrometer long with an average diameter of 60 nm. These nanowires were synthesized in mass production via thermal evaporation technique. Purity of these $ZnSnO_3$ nanowires is reported to be very good. Sensors realized from these $ZnSnO_3$ nanowires were very sensitive to ethanol gases. It is reported that the sensitivity was sufficiently high even at 1 ppm ethanol exposure. Both the response and recovery time of these sensors were very short and measured about 1 s. Their experimental results strongly suggested that $ZnSnO_3$ nanowires could be an excellent candidate for applications in gas sensors at the industry level and also in the field of alcohol vapor sensing. Extremely high oxygen sensing is also realized from individual nanowires with abundant grain boundaries [264].

In the field of agricultural and biotechnological processes as well as air-conditioning systems or monitoring of exhaust gases, there is an increasing demand on CO_2 sensors. It is pointed out by Keller et al. [27] that the nanocrystalline materials, such as $BaTiO_3$, increase the performance of CO_2 sensors. They have proposed a new generation of low-cost CO_2 sensors using hybrid thick-film technology. The innovation presented by them is the use of a mixture of nanocrystalline $BaTiO_3$ and other oxides with grain sizes below 50 nm. The synthesized nanopowders were found to be 25 nm in diameter, as shown in Figure 7.105a, and the powders were actually mixtures of CuO, TiO_2, and $BaCO_3$ besides $BaTiO_3$. After heat treatment at 750°C, no major modification or grain growth was observed in nanopowders as indicated in Figure 7.105b. All nanocrystalline sensors fabricated by these materials have shown a long-term stability.

Nanocrystalline $BaTiO_3$ material has better humidity sensing properties than ceramic $BaTiO_3$, such as lower resistance and high sensitivity. However, from the application point of view, one needs to further reduce the resistance of the resistive humidity sensor from 10^6 to 10^3, to minimize the humidity hysteresis and to increase the sensitivity and repeatability [265]. Efforts are continuing to improve the properties of the nanocrystalline $BaTiO_3$ humidity sensor by doping a certain amount of impurities into $BaTiO_3$, and mixing some polymer humidity sensing materials with $BaTiO_3$ to fabricate the humidity sensors of composite materials.

By using solvothermal reaction between metal alkoxides and benzyl alcohol enables the synthesis of a variety of $BaTiO_3$ oxides [35]. The as-synthesized nanoparticles were crystalline and exhibited uniform particle morphology with a narrow size distribution within one system. One such particle

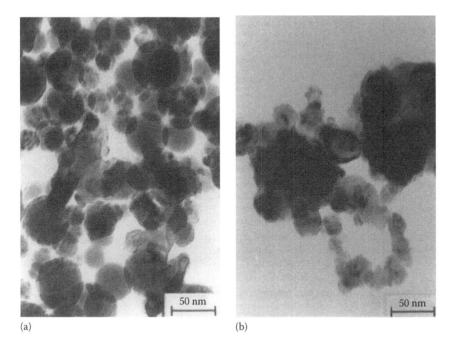

(a)   (b)

FIGURE 7.105

TEM micrographs of mix-BaTiO$_3$ nanopowder (a) as prepared and (b) after heat treatment at 750°C. (Reprinted from *Sens. Actuators*, B57, Keller, P., Ferkel, H., Zweiacker, K., Naser, J., Meyer, J.-U., and Riehemann, W., The application of nanocrystalline BaTiO$_3$-composite films as CO$_2$-sensing layers, 39–46, Copyright 1999, with permission from Elsevier Science S.A.)

morphology of BaTiO$_3$ is shown in Figure 7.106, with individual nanoparticles with diameters of 4–5 nm and uniform in size and shape. The image at higher magnification gives evidence for the high crystallinity as shown in figure inset.

By using electrospinning technique, nanofibers of complex oxides were synthesized, particularly the BaTiO$_3$ phase, by Yuh et al. [192]. The phase formation and morphology of these nanofibers was investigated as a function of heat treatment conditions. Figure 7.107a shows the SEM images of as-synthesized and dried samples at 120°C, and Figure 7.107b presents the SEM images of BaTiO$_3$ nanofibers that were heat treated at 550°C for 12 h. By this process it was observed a reduction of 60% of its diameter because of polyvinyl pyrrolidone (PVP) burnout without losing its smooth surface morphology and fiber continuity. At this stage, fiber diameters ranged from 60 to 130 nm. The Figure 7.107c shows the samples treated for 16 h at 600°C temperature, and Figure 7.107d for 16 h treatment at 750°C. Fully crystallized BaTiO$_3$ nanofibers with the perovskite structure are obtained after annealing the samples at 750°C. Tetragonal crystal structure of the fibers is indicated by XRD peak splitting and it was further confirmed by Raman Spectroscopy technique. Furthermore, the advancement in heat treatment of

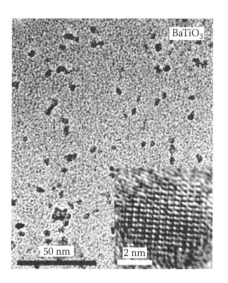

FIGURE 7.106
Transmission electron micrograph of BaTiO$_3$ nanoparticles together with HRTEM image of single particle as inset. (Reprinted from *Prog. Solid State Chem.*, 33, Niederberger, M., Garnweitner, G., Pinna, N., and Neri, G., Non-aqueous routes to crystalline metal oxide nanoparticles: Formation mechanisms and applications, 59–70, Copyright 2005, with permission from Elsevier Ltd.)

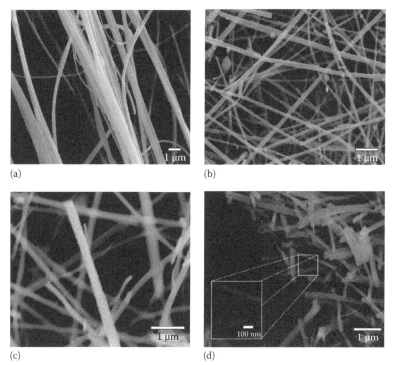

FIGURE 7.107
SEM images of BaTiO$_3$ nanofibers under different conditions: (a) as-synthesized and dried at 120°C, (b) heat treated at 550°C for 12 h, (c) 600°C for 16 h, and (d) 750°C for 16 h. Lower left inset shows the magnified image of a single nanofiber. (Reprinted from *Phys. E Low Dimens. Syst. Nanostruct.*, 37, Yuh, J., Perez, L., Sigmund, W.M., and Nino, J.C., Electrospinning of complex oxide nanofibers, 254–259, Copyright 2007, with permission from Elsevier B.V.)

the electrospun fibers yields single crystalline $BaTiO_3$ nanofibers with 50 nm in diameter and lengths up to 1 mm.

Studies of Kosacki and Anderson [266] on electrical properties of nano-structured $SrCe_{0.95}Yb_{0.05}O_3$, with grain size of the order 7–70 nm, indicated that grain boundary conductivity is related to their microstructure. These nanostructured films possess greatly enhanced conductivity and faster kinetic reactions with ambient atmosphere due to the grain boundary and surface controlled diffusion rates than microcrystalline specimens of the same composition. In these sensors, the electrical conductivity was measured by impedance spectroscopy. By using the sensors, it is possible to detect a wide range, of the order of 3%–100%, but needs relatively high temperature to operate them.

Nanoparticles of nickel ferrite, which show n-type semiconductor behavior, were partly replaced with cobalt and manganese in place of nickel and iron to improve the sensitivity of the gas-sensing properties, particularly for LPG detection. It was shown, by Satyanarayana et al. [16] that ferrite compounds incorporated with palladium demonstrated exceptionally improved gas-sensing characteristics and were found to be selective to detect reducing gases, such as CH_4, CO, and C_2H_5OH, at low temperature ranges. The SEM morphology of the sample calcined in air at 500°C is shown in Figure 7.108a. Gas-sensing measurements of these compounds show that the sensitivity has increased with an increase in the operating temperature. These enhanced properties are shown in Figure 7.108b. It is indicated that the presence of oxygen vacancies in the semiconducting oxide ferrite is principally responsible for this detection.

Transition-metal oxides constitute an attractive class of inorganic solids due to a variety of structure, properties, and phenomenon. Among the transition-metal oxides, spinel-type AB_2O_4, where element A and B denote divalent and trivalent metallic cations, respectively, are very interesting materials with improved reactivity than the corresponding single oxides. In particular, the cobalt-containing spinel oxides MCo_2O_4 (M = Ni, Cu, Zn, Mg, Mn, Cd, etc.) are technologically intriguing materials and have found many applications in the areas such as chemical sensors, electrode material, electrocatalyst, and pigment. To date, three morphological MCo_2O_4 (M = Ni, Cu, Zn) nanostructures of nanoparticles, nanofibers, and nanofilms have been prepared by various techniques. Some of these oxides are good candidates for gas-sensing applications.

Hu et al. [267] used high-energy ball milling technique to synthesize $SrTiO_3$ nanosized powders. $SrTiO_3$ is a very important material for oxygen sensors. It has attracted much attention because of their low-cost and strong stability in thermal and chemical atmospheres. Most research works focus on bulk-conduction-type in $SrTiO_3$ with high operating temperatures (700°C–1000°C), and $SrTiO_3$ sensing materials are produced by the conventional high temperature solid-state reaction method because of its high melting point (of the order of 2080°C). At low oxygen concentrations the relative resistance

(a)

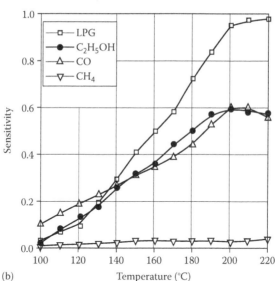

(b)

FIGURE 7.108
(a) Scanning electron micrograph of the sample $Ni_{1-x}Co_xMn_xFe_{2-x}O_4$ calcined at 500°C in air, and
(b) gas-sensing characteristics of the sample at different temperatures. (Reprinted from *Sens. Actuators*, B89, Satyanarayana, L., Reddy, K.M., and Manorama, S.V., Synthesis of nanocrystalline $Ni_{1-x}Co_xMn_xFe_{2-x}O_4$: A material for liquefied petroleum gas sensing, 62–67, Copyright 2003, with permission from Elsevier Science B.V.)

value increases rapidly with the increase in the O_2 concentration, however, at higher concentrations, the increase in resistance value becomes more gradual. The highest relative resistance for the nanostructured synthesized $SrTiO_3$ sensor devices to 20% oxygen is found to be very high with operating temperature at 700°C–800°C. The detected range is 1%–20% oxygen, which

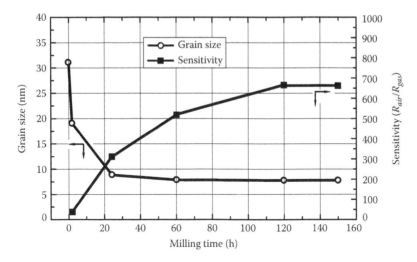

FIGURE 7.109
Correlation between grain size and sensitivity for different milling times for $xSnO_2$-$(1-x)$ α-Fe_2O_3. (Reprinted from *Sens. Actuators*, B65, Tan, O.K., Zhu, W., Yan, Q., and Kong, L.B., Size effect and gas sensing characteristics of nanocrystalline $xSnO_2$-$(1-x)$α-Fe_2O_3 ethanol sensors, 361–365, Copyright 2000, with permission from Elsevier Science S.A.)

is a very wide range possible for a single type of sensor material. Recently, a nanostructured *p*-type oxide, $SrTiO_3$, was reported as an oxygen sensor operating at a temperature of 40°C [227]. Regarding *n*-type semiconducting oxides, only a few reports are found in the literature.

Tan et al. [229] prepared nonequilibrium nanocrystalline $xSnO_2$-$(1-x)$ α-Fe_2O_3 powders by using mechanical alloying technique. Ethanol sensors were fabricated by using screen printing technology. It is observed that the particle size of the nanopowders drastically reduced to 10 nm size after 24 h of milling and remains about the same up to 150 h of milling as shown in Figure 7.109. The decrease in average grain size has also been confirmed by TEM analysis. The figure also shows the gas sensitivity to 1000 ppm ethanol in air obtained for the sample with $x = 0$ mol%, operating at temperature of 257°C. It exhibits an inverse relationship between gas sensitivity and milling time. As the milling time increases, the grain size decreases. This would greatly increase the surface activity since the surface area has increased. It is believed that the sensitivity is directly linked to the surface activity. The enormous oxygen dangling bonds at the particle surfaces have given rise to the increase in sensitivity.

Zhang et al. [201] prepared nanotubes by porous AAO-template method. Figure 7.110 shows the SEM images of the $NiCo_2O_4$ products at different magnifications. Figure 7.110a is a typical SEM image at a relatively low magnification, showing the overall view of the $NiCo_2O_4$ nanotubes. From which, it can be seen that a large quantity of nanotube-bundle filaments and a small quantity of scattering tube filaments were obtained. The length of the nanotube

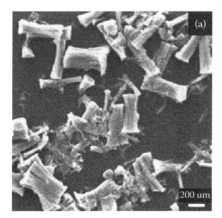 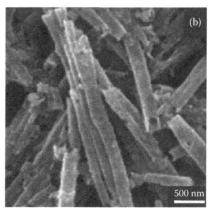

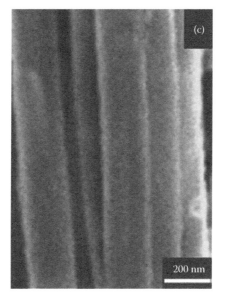

FIGURE 7.110

Typical SEM images of as-synthesized $NiCo_2O_4$ nanotubes: (a) an overall view at a low magnification, (b) the scattering nanotubes, and (c) the nanotube-bundle wall at a relatively high magnification. (Reprinted from *Sens. Actuators*, B114, Zhang, G.-Y., Guo, B., and Chen, J., MCo_2O_4 (M = Ni, Cu, Zn) nanotubes: Template synthesis and application in gas sensors, 402–409, Copyright 2006, with permission from Elsevier B.V.)

bundles is approximately 50–60 μm, being consistent very well with the thickness of the template membrane employed in the synthesis. Figure 7.110b displays the SEM image of the scattering nanotubes at a high magnification. Apparently, the scattering tubes are much shorter than the bundle tubes, indicating that the long-bundle tubes were broken into shorter ones due to the further dissolution of the AAO membrane. The open-ended tips can be seen from some tubes perpendicular to the image. Figure 7.110c shows the

SEM image of one tube bundle at a higher magnification. The much clearer image shows that the $NiCo_2O_4$ nanotubes are highly uniform cylinders and arranged roughly parallel to each other. The average outer diameters of the nanotubes are about 200 nm, which are almost the same as the pore diameters of the template used. Figure 7.111 shows the gas response of the as-prepared $NiCo_2O_4$ nanotubes to CH_3COOH of different concentrations at 300°C in comparison with the corresponding nanoparticles, respectively. From the figure, it is apparent that the response of the $NiCo_2O_4$ nanotubes, which increases from about 3 times at 50 ppm up to 35 times at 400 ppm, is comparable with that of the $NiCo_2O_4$ nanoparticles. Similarly, $CuCo_2O_4$ also shows different response to SO_2 gas. Figure 7.111a shows the concentration-dependent response of the sensors made by $CuCo_2O_4$ nanoparticles (circles) and nanotubes (squares) to SO_2, and Figure 7.111b shows the $NiCo_2O_4$ nanoparticles (circles) and nanotubes (squares) to CH_3COOH at 300°C. Nanoparticles have limited surface area, whereas nanotubes have larger surface area and exhibit more sensitivity. This brings out clearly the difference between the large surface area and limited surface area and the relationship with sensor sensitivity factors.

Liu et al. [32] have demonstrated the effect of platinum doping to $CdSnO_3$. The sensor responses for undoped and doped with different content of platinum as a function of operating temperature toward 50 ppm ethanol gas are presented in Figure 7.112a. The sensitivity goes through a maximum at around 200°C for each $CdSnO_3$ sensor and maximum sensitivity to ethanol gas was measured up to 68.2 obtained by 1.5 at.% platinum doped sample. It is evident that platinum is beneficial for the sensitivity of $CdSnO_3$ to ethanol gas, as the noble metal usually acts as an efficient catalyst, and the sensitivity increases with increment in concentration of platinum up to 1.5 at.%. The selectivity of same samples to various reducing gases like LPG, CH_4, CO, butane (C_4H_{10}) and gasoline was also measured. The details are shown in Figure 7.112b. The sensitivity to 50 ppm ethanol gas at 200°C is as high as 68.2, while it reaches only 2.0, 2.5, 3.6, 9.8, and 10.8 for 500 ppm CH_4, C_4H_{10}, CO, LPG, and 50 ppm gasoline, respectively. It is obvious that the developed ethanol gas sensor possesses an excellent selectivity. It is clearly observed that, in agreement with the structural characterization, resistance variations are better explained through changes in barrier height produced by the presence of the additives at the surface, than through doping variations. On the other hand, it is reasonable to think that the used impregnation procedure should not substantially vary the bulk doping level.

Table A.22 shows the mixed oxide composition and the dopants used. Their preparation techniques along with grain sizes are indicated. Mixed oxides option gives a scope for different gaseous species selection and it is possible to identify material compositions so that the synthesized composition is suitable to a group of gaseous species. One can tailor-make such compositions suitably. The response time of these materials is relatively large and these materials may not be highly suitable for quickly changing environments.

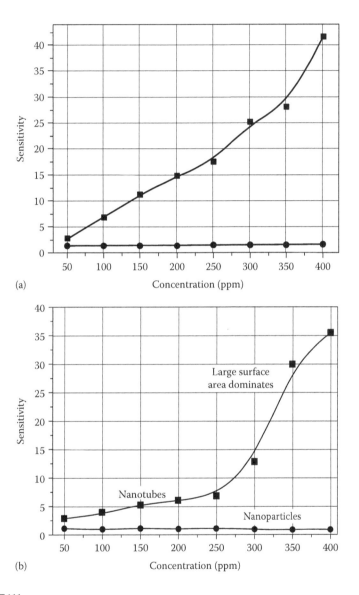

(a)

(b)

FIGURE 7.111
Concentration-dependent response of the sensors made by (a) CuCo$_2$O$_4$ nanoparticles (circles) and nanotubes (squares) to SO$_2$. (b) NiCo$_2$O$_4$ nanoparticles (circles) and nanotubes (squares) to CH$_3$COOH at 300°C. Nanoparticles have limited surface area, whereas nanotubes have larger surface area and exhibit more sensitivity. (Reprinted from *Sens. Actuators*, B114, Zhang, G.-Y., Guo, B., and Chen, J., MCo$_2$O$_4$ (M = Ni, Cu, Zn) nanotubes: Template synthesis and application in gas sensors, 402–409, Copyright 2006, with permission from Elsevier B.V.)

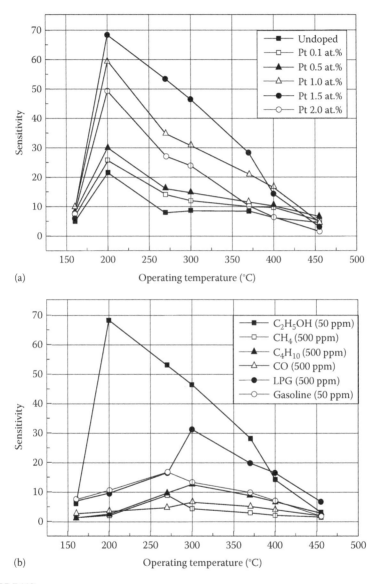

FIGURE 7.112
(a) Sensitivity to 50 ppm ethanol versus operating temperature for undoped CdSnO₃ sensor and for CdSnO₃ samples activated by Pt with the atomic weight percentage ranging from 0.1 to 2.0 at.%. (b) Sensitivity of sensors doped with 1.5 at.% Pt to different gases at different operating temperatures. (Reprinted from *Anal. Chim. Acta*, 527, Liu, Y.-L., Xing, Y., Yang, H.-F., Liu, Z.-M., Yang, Y., Shen, G.-L., and Yu, R.-Q., Ethanol gas sensing properties of nano-crystalline cadmium stannate thick films doped with Pt, 21–26, Copyright 2004, with permission from Elsevier B.V.)

Though the thin-film approach is better for the development of gas-sensing devices, for further optimization of the sensing properties of these films it is necessary to develop different novel approaches and active devices that will further increase the sensitivity, selectivity, and decrease response and recovery times and also the necessary working temperatures of the sensing elements. It is also necessary to identify those specific devices for better utilization.

8

Active Devices Based on Nanostructures

The method of fabricating nanoelectronics is vastly different from the conventional microelectronics processing technology. The devices are too small, a mere few nanometers large, and an estimated 10^{12} per can be integrated per square centimeter area when compared to 10^{10} per using MOSFET devices, projected for the year 2018 [271]. It should be noted that this level of integration will be hard to attain because of those nanolevel dimensions; the arbitrary patterning of circuits will probably not be feasible. Because of the difference in fabrication techniques, nanowires (NWs) and nanobelts are more attractive where the conduction properties can be more tightly controlled. Thus, they have more active device potential than the nanotubes. These semiconducting metal-oxide NWs can be used as an active device and also as interconnecting wires to carry electrical signals [272]. In addition to improved electrostatics, the one-dimensional (1-D) transistors provide novel characteristics due to quantum confinement. A recent study of ballistic transistors [273] shows, however, that except for differences in electrostatics, device performance metrics, such as the injection velocity and the intrinsic device delay, are similar for NW transistors and two-dimensional (2-D) planar transistors.

8.1 Diodes and Schottky Diodes

Ultraviolet (UV)-light irradiation of the nanobelt diode of SnO_2 in air results in a significant increase of the conductivity of the single nanobelt structure. Light with a wavelength of 350 nm, equivalent to $E_\lambda = 3.54$ eV, was used for this purpose. This energy was in excess of the direct bandgap of SnO_2. The increase in the conductivity results from photogeneration of electron-hole pairs as well as doping by UV light–induced surface desorption [60]. The current versus voltage (I–V) characteristic curves of a diode, made using a single SnO_2 nanobelt, showed variable response with and without UV radiation.

Pt/ZnO Schottky diodes fabricated on bulk ZnO show changes in forward current of 0.3 mA at a forward bias of 0.5 V or alternatively a change of 50 mV bias at a fixed forward current of 8 mA when 5 ppm of H_2 is introduced into an N_2 ambient at 25°C [186]. The rectifying current–voltage characteristic shows a nonreversible collapse to simple ohmic behavior when as little as 50 ppm of

H_2 gas is present in the N_2 ambient. It had been observed that at higher temperatures, the recovery is thermally activated with an activation energy value of ~0.25 eV. This observation suggests that introduction of hydrogen introduces shallow donors into the ZnO and is a contributor to the change in current of the diodes. Such changes were also observed with diluted vapors of ethylene in nitrogen ambient. UV radiation also affects the electrical properties of these diodes almost in the same way. The *I–V* characteristics at 150°C of the Pt/ZnO diode both in pure N_2 and in ambients containing various concentrations of C_2H_4 are shown in Figure 8.1. At a given forward or reverse bias, the current increases upon introduction of the C_2H_4, through a lowering of the effective barrier height. One of the main mechanisms is once again the catalytic decomposition of the C_2H_4 on the Pt metallization, followed by diffusion to the underlying interface with the ZnO. These changes are reproducible and highly reliable ZnO-based Schottky diodes are expected in near future to measure such fine concentration values of hydrogen and also the ethylene vapors. So far, no sensing material is identified that detects this important volatile organic compound.

Low-temperature hydrogen sensors based on gold nanoclusters and Schottky contacts on ZnO films was reported by Pandis et al. [161]. These films were deposited by using pulsed laser deposition technique on Si/SiO$_2$ substrates. Unintentionally doped *n*-type ZnO thin films were functionalized as the H_2 gas sensors by incorporating gold nanoclusters on the surface and thus developing

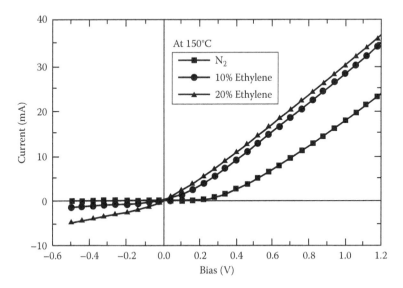

FIGURE 8.1

Current–voltage characteristics of Pt/ZnO Schottky diodes measured at different gas concentration values of ethylene in pure and mixed nitrogen ambients at 150°C. (Reprinted from *Mater. Sci. Eng.*, R47, Heo, Y.W., Norton, D.P., Tien, L.C., Kwon, Y., Kang, B.S., Ren, F., Pearton, S.J., and LaRoche, J.R., ZnO nanowire growth and devices, 1–47, Copyright 2004, with permission from Elsevier.)

Au/ZnO Schottky diodes. The influence of the catalytic action of the gold nano-clusters on the sensing properties of the devices was examined and found to provide faster response times at a reduced working temperature of 150°C. *I–V* characteristics of these Schottky diodes showed a remarkable increase in the current and were controlled by the applied bias voltage. The characteristics were also examined under exposure to 3% H_2 mixture with air at working temperatures between 30°C and 90°C and the devices showed faster response to H_2.

Application of Schottky diodes is advantageous for gas-sensing purposes due to the simple electrical circuitry required to operate them. The simplest form of SiC Schottky diode-based gas sensor is one that consists of a catalytic metal deposited on the semiconducting SiC. Trinchi et al. [269] presented propene (C_3H_6) gas sensor utilizing similar configuration of Pt/Ga_2O_3-ZnO/SiC Schottky diodes. The gases, particularly the hydrogen and hydrocarbons, dissociate on the catalytic metal surface and diffuse through the metal/semiconductor interface. This results in a dipole layer property that changes the Schottky diode electrical properties, by lowering the barrier height, in proportion to the concentration of the gaseous species. It is expected that further investigation and layer modifications of these Schottky diodes may provide greater understanding of the gas-sensing mechanism of these devices.

8.2 Field-Effect Transistors

For the development of Si-based field-effect transistor (FET) sensors, thin film of a gas-sensitive material at a temperature below the range of 200°C is a prerequisite [92] for its operation. Different designs of FET structures are being reported to develop such FET-based gas sensors for industrial and environmental applications. In particular, capacitively controlled FET (CCFET) and hybrid suspended gate FET (HSGFET) structures allow the possibility of fabricating a sensitive layer separately before integrating it with the device as its gate electrode. This approach provides an opportunity to investigate gas-sensing properties based on work function change of a wide variety of materials, even prior to their implementation in a FET device. Currently, research efforts are concentrated in the search of an appropriate gas-sensitive material that is suitable to realize a Si FET sensor.

8.3 FET-Based Gas Sensors

Wang et al. [60,274] first reported the FETs fabricated from single crystalline SnO_2 and ZnO nanobelts. These devices have switching ratios as large as six orders of magnitude, high conductivity value of 15 (Ω cm)$^{-1}$, and electron mobility as large as 125 cm^2/Vs. Furthermore, SnO_2 nanobelts doped with surface oxygen vacancies by annealing them in reduced oxygen environments increased the conductivity and drastically decreased the gate threshold voltage, indicating the feasibility of tuning the device characteristics by controlling adsorbed oxygen species. Short-channel effects are also reported in SnO_2 nanobelts showing the inability of the gate electrode to modulate source-drain channel conductivity in nanobelts shorter than about 500 nm. The scientific group has demonstrated the fabrication of nanoscale FETs using SnO_2 and ZnO nanobelts as the FET channels, analyzed their *I–V* characteristics, and demonstrated the sensitivity of the SnO_2 nanobelt conductance to gas exposure.

Zhang et al. [83] have reported the transport studies on FETs made of individual In_2O_3 NWs, which were synthesized via a novel CVD method. These devices exhibited pronounced gate-dependence and well-defined linear and saturation regimes. Thermal emission was found to be the dominating transport mechanism. Oxygen molecules adsorbed on the NW surface was found to have a profound effect on the transistor performance, as manifested by a 10-fold increase in conduction, one order of magnitude increase in the transistor on/off ratio and a significant shift of the device threshold voltage. As shown in Figure 8.2, the devices were fabricated on $Si–SiO_2$ platform and channel length between the two electrodes is 2 μm. Highly doped Si substrate was used as a back gate. The figure shows a typical FET $I–V_{ds}$ characteristic curve of the device at room temperature. Six $I–V_{ds}$ curves at V_g values of 15,

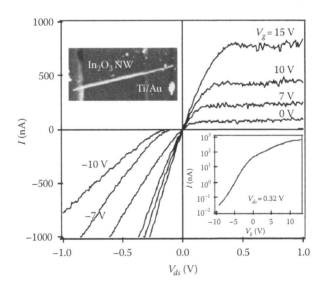

FIGURE 8.2
Gate-dependent *I–V* curves recorded at room temperature for In_2O_3 NW. The lower inset shows the current versus gate voltage at $V_{ds} = 0.32$ V. The gate modulated the current by five orders of magnitude. The upper left inset is an AFM image of the In_2O_3 NW between two electrodes. (Reprinted with permission from Zhang, D., Li, C., Han, S., Liu, X., Tang, T., Jin, W., and Zhou, C., Electronic transport studies of single-crystalline In_2O_3 nanowires, *Appl. Phys. Lett.*, 82, 112–114. Copyright 2003, American Institute of Physics.)

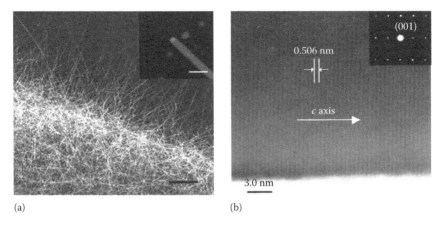

FIGURE 8.3
(a) FE-SEM image of as-grown ZnO NWs, scale bar is 20 µm. Inset image shows a single ZnO NW terminated with a gold nanoparticle, scale bar is 100 nm. (b) HRTEM image and diffraction pattern of a ZnO NW. (Reprinted with permission from Fan, Z., Wang, D., Chang, P.-C., Tseng, W.-Y., and Lu, J.G., ZnO nanowire field-effect transistor and oxygen sensing property, *Appl. Phys. Lett.*, 85, 5923–5925. Copyright 2004, American Institute of Physics.)

10, 7, 0, −7, and −10 V. With the gate voltage varying from +15 to −10 V, the conductance of the NW was gradually suppressed. The results indicate that these In_2O_3 NWs have great potential to be used as nanoelectronic building blocks and ultrasensitive chemical sensors.

Fan et al. [173] synthesized *n*-type single crystal ZnO NWs by a vapor-trapping CVD method, seeded with gold nanoparticles. Figure 8.3a shows the FE-SEM image of these as-grown NWs. These NWs, terminated with a gold nanoparticle, were formed with an average diameter of 60 nm and lengths varied up to several tens of microns. The inset clearly shows the gold nanoparticle. Figure 8.3b shows the HRTEM image and diffraction pattern of a ZnO NW. Electrical transport properties were investigated for these NWs configured as FETs. The charge-carrier concentration and electron mobility parameters were estimated to be $\sim 10^7$ cm^{-1} and ~ 17 cm^2/V·s, respectively. A contact barrier to the fabricated FET devices was characterized, and thermionic emission was found to dominate the current transport. Oxygen adsorption on to the NW surface was shown to have considerable effect on the measured conductance, and an oxygen-sensing study showed that the sensitivity depends on NW diameter as well as on the applied gate voltage. These results indicate a new way of utilizing these ZnO NWs as building blocks for futuristic nanoscale electronics and also for the development of miniaturized chemical sensing devices.

Li et al. [174] synthesized ZnO NWs by thermal evaporation of zinc powders under oxygen ambient at a temperature of 1100°C. The diameter of these NWs varied between 40 and 60 nm as shown in Figure 8.4. Figure 8.4a shows the SEM image of the ZnO NWs, and Figure 8.4b shows the

(a)

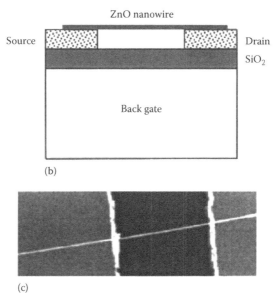

(b)

(c)

FIGURE 8.4
(a) The SEM image of the ZnO NWs. (b) Schematic illustration of the ZnO NW transistor; a single ZnO NW connects the two electrodes (source and drain). The substrate is used as the back gate. (c) The SEM image of the transistors, the separation between the two electrodes is about 1 μm. (Reprinted with permission from Li, Q.H., Liang, Y.X., Wan, Q., and Wang, T.H., Oxygen sensing characteristics of individual ZnO nanowire transistors, *Appl. Phys. Lett.*, 85, 6389–6391. Copyright 2004, American Institute of Physics.)

schematic illustration of the ZnO NW transistor where a single ZnO NW connects the two electrodes of source and drain. Here, the substrate is used as the back gate. Figure 8.4c shows the SEM image of the transistors and the separation between the two electrodes is about 1 μm. By using the conventional photolithography technique gold electrodes were defined to study these NWs. By using dispersed NWs they were defined on the electrodes as bridge configuration structures as shown in the figure. Individual ZnO NW transistors were thus fabricated were characterized in a vacuum chamber at different oxygen pressures. These transistors exhibited high sensitivity to oxygen gas, which led to a change of the source-drain current and a shift in the threshold voltage. These results implied that the sensing properties of the transistors are related to the trapping and releasing of the carriers in the NWs. These devices also showed quick response to UV radiation. This also indicates a possible application of these NWs in the application field of UV radiation detectors.

Large variety of ZnO 1-D structures have been demonstrated for device applications [186]. The large surface area of these nanorods and biosafe characteristics of this ZnO material makes them attractive for gas and chemical sensing and also for biomedical applications. The initial reports on electrical characteristics show a pronounced sensitivity in electrical conductivity to UV illumination and the presence of oxygen in the measurement ambient. ZnO can be grown at low temperatures on cheap substrates, such as glass, makes it attractive for transparent electronics. A single NW Schottky diode was fabricated by using e-beam lithography technique to pattern sputtered Al/Pt/Au electrodes contacting both ends of the NWs. The separation of the electrodes was ~3 μm. E-beam evaporated Pt/Au was used as the gate metallization by patterning a 1 μm wide strip orthogonal to the NW. A scanning electron micrograph of the completed device is shown in Figure 8.5. Gold wires were bonded to the contact pads for current–voltage measurements. These measurements were carried out at room temperature (25°C). In some cases, the diodes were illuminated with above band edge UV radiation during the measurement. Figure 8.6 shows NW FET configuration and well-defined source, gate, and drain regions. MOSFET device configurations are shown in Figure 8.7. Figure 8.7a shows the SEM micrograph of ZnO MOSFET structures and Figure 8.7b shows the close-up view of the actual nanorod and its contacts. These devices exhibited typical *I–V* characteristics similar to standard electronic devices. These NW-based ZnO active devices are expected to play a key role in near future.

NWs can be configured in a FET structure with a three-terminal configuration and they can be tested in single wire or multiple wire FETs. In such configurations, the Fermilevel within the band gap of the NW can be varied and used to control surface process electronically. The NW acts as a conductive channel that joins source and drain electrodes. Tuning of the metal-oxide properties in a FET configuration has been reported and these devices showed good switching ratios between ON and OFF states, as shown in Figure 8.8. Figure 8.8a shows the scanning electron micrograph of

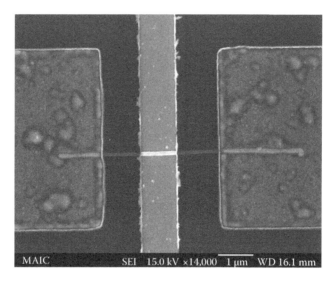

FIGURE 8.5
SEM micrograph of ZnO NW Schottky diode. (Reprinted from *Mater. Sci. Eng.*, R47, Heo, Y.W., Norton, D.P., Tien, L.C., Kwon, Y., Kang, B.S., Ren, F., Pearton, S.J., and LaRoche, J.R., ZnO nanowire growth and devices, 1–47, Copyright 2004, with permission from Elsevier.)

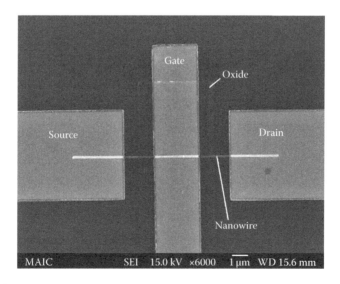

FIGURE 8.6
SEM micrograph of fabricated FET. (Reprinted from *Mater. Sci. Eng.*, R47, Heo, Y.W., Norton, D.P., Tien, L.C., Kwon, Y., Kang, B.S., Ren, F., Pearton, S.J., and LaRoche, J.R., ZnO nanowire growth and devices, 1–47, Copyright 2004, with permission from Elsevier.)

(a)

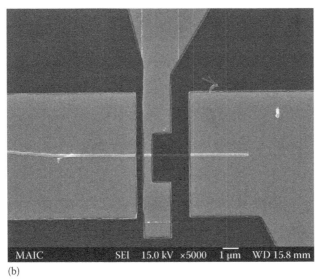

(b)

FIGURE 8.7
(a) SEM micrograph of ZnO MOSFET structures. (b) Close-up view of the actual nanorod and its contacts. (Reprinted from *Mater. Sci. Eng.*, R47, Heo, Y.W., Norton, D.P., Tien, L.C., Kwon, Y., Kang, B.S., Ren, F., Pearton, S.J., and LaRoche, J.R., ZnO nanowire growth and devices, 1–47, Copyright 2004, with permission from Elsevier.)

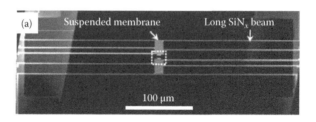

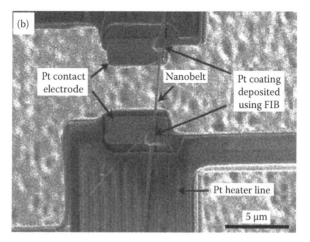

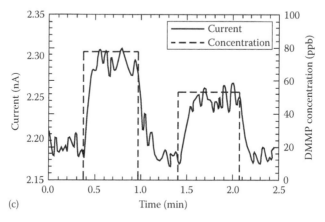

FIGURE 8.8

(a) Scanning electron micrograph of the microheater device. (b) An enlarged image showing the region inside the dashed rectangle of (a). (c) Response of the as-assembled SnO_2 nanobelt sensor to 78 and 53 ppb DMMP balanced with air when the nanobelt temperature was 500°C. The voltage applied to the nanobelt was 1.5 V. (Reprinted with permission from Yu, C., Hao, Q., Saha, S., Shi, L., Kong, X., and Wang, Z.L., Integration of metal oxide nanobelts with microsystems for nerve agent detection, *Appl. Phys. Lett.*, 86, 063101. Copyright 2005, American Institute of Physics.)

the microheater device, and Figure 8.8b shows an enlarged image showing the region inside the dashed rectangle of (a). Figure 8.8c shows the response of the as-assembled SnO_2 nanobelt sensor to 78 and 53 ppb dimethyl methylphosphonate (DMMP) balanced with air when the nanobelt temperature was 500°C. The voltage applied to this nanobelt was 1.5 V. The channel conductance and the threshold values are reported to be sensitive to the surrounding atmospheres.

Ng et al. [181] have demonstrated the importance of substrate engineering to grow highly regular and vertically aligned NW arrays, and a unique direct integration of the NWs using the present semiconductor processing technology, bridging the gap between microtechnology and the nanotechnology. The device fabrication approach has allowed a lithography-free means of defining the vertical channel length by chemical mechanical polishing (CMP) and reduced the footprint of the devices since the drain, source, and channel of FET are stacked on top of each other in a vertical way. Figure 8.9a shows FE-SEM micrograph of cladded ZnO NW and a vertically aligned ZnO NW projecting from a dome-like ZnO buffer layer. The inset shows its top-view FE-SEM image, revealing that the hexagonal geometry of the buffer layer. Figure 8.9b shows the same NW cladding with Cr (10 nm)/SiO_2 (20 nm) layer. Figure 8.9c shows the top view FE-SEM image of the cladded ZnO nanostructure prior to recess formation and deposition of the top Cr drain electrode. The surround-gate oxide and the Cr gate electrodes are clearly shown here. Figure 8.9d illustrates the three-dimensional schematic on the critical components of vertical surround-gate FET (VSG-FET). These VSG-FETs were fabricated with cell size featuring at least a 10% smaller footprint than the current state-of-the-art MOSFET technology with a potential of achieving tera-level ultrahigh packing density and radiation-hard electronic devices. Figure 8.9 shows the FE-SEM micrographs of cladded ZnO NWs where a vertically aligned ZnO NW is projecting from a "dome"-like ZnO buffer layer revealing the hexagonal geometry of the buffer layer. The same NW cladded with Cr (10 nm)/SiO_2 (20 nm) layers and the corresponding top-view FE-SEM image showing broadening of the diameter of the cladded ZnO nanostructure where the hexagonal base is clearly shown. The surround-gate oxide and Cr gate electrode are also clearly observed, showing the highly conformal CVD and ion-beam deposition processes.

ZnO nanobelt FETs are sensitive to UV light. Both photo generation of electron-hole pairs and doping by UV induced surface desorption contribute to the conductivity. In these nanobelts, ZnO has direct bandgap and two processes are observed to contribute to photoconductivity. The first process involves the photo generation of electron-hole pairs, whereas the second process most likely involves chemical desorption from the ZnO surface by exposure to UV light.

ZnO NW depletion mode MOSFETs show excellent pinch-off and saturation characteristics and a strong UV photo response [186]. These devices

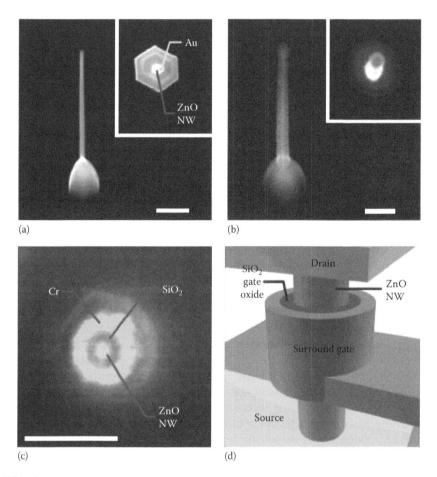

FIGURE 8.9
FE-SEM micrograph of cladded ZnO NW. (a) A vertically aligned ZnO NW projecting from a dome-like ZnO buffer layer. The inset shows its top-view FE-SEM image, revealing the hexagonal geometry of the buffer layer. (b) The same NW cladding with Cr (10 nm)/SiO$_2$ (20 nm) layer. Corresponding top-view FE-SEM image shows broadening of the diameter of the cladded ZnO nanostructure. (c) Top-view FE-SEM image of the cladded ZnO nanostructure prior to recess formation and deposition of the top Cr drain electrode. The surround-gate oxide and Cr gate electrode can be clearly observed, showing the highly conformal CVD and ion-beam deposition processes. (d) A 3-D schematic illustrating the critical components of VSG-FET. (Scale bar: 200 nm for (a), (b), and (c). (Reprinted with permission from Ng, H.T., Han, J., Yamada, T., Nguyen, P., Chen, Y.P., and Meyyappan, M., Single crystal nanowire vertical surround-gate field-effect transistor, *Nano Lett.*, 4, 1247–1252. Copyright 2004, American Chemical Society.)

look promising for transparent transistor applications requiring low leakage current and indicate that the NWs can be grown and transferred to another substrate without major degradation of their electrical transport properties.

The potential application of ZnO to electronic and optoelectronic devices is due to large excitation binding energy of 60 meV [275]. The relatively low electron mobility values may restrict its usage. The other problem is about *p*-type ZnO material that is necessary to produce light emitters. On the electronic side, unipolar devices can be considered without *p*-type doping for this material. An atomic force microscopy image of the FET is also reported [60]. The principle of this device is that controlling the gate voltage controls the current flowing from the source to the drain.

ZnO nanorods are promising candidates for detecting extremely low concentrations of CO and H_2S gaseous species. The ratio of the electrical resistance of these rods in air to that in 0.05 ppm H_2S was measured to be 1.7 at room temperature [276]. The selectivity is achieved by applying different voltages to the gate of a NW FET or by performing measurements at different temperatures since different gas molecules have different activation energies. Figure 8.10 [276,277] shows an example of such selectivity for NH_3 and NO_2 gas molecules, where the refresh (erase) voltages, negative gate voltages required for electrical desorption of adsorbed gas molecules, for the two gas molecules are significantly different making it possible to distinguish different gas species. It is further reported [277] that the detection sensitivity can be tuned by the back gate potential and large negative gating could significantly expedite the desorption process at room temperature. Furthermore, the gate potential variation induced time-dependent behavior demonstrated a potential gas-distinguishing mechanism.

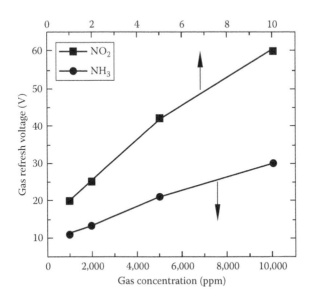

FIGURE 8.10
Gate refresh voltages of a ZnO NW FET for NH_3 and NO_2 gas molecules as a function of gas concentration. (Reproduced from Özgür, Ü., Hofstetter, D., and Morkoç, H., ZnO devices and applications: A review of current status and future prospects, *Proc. IEEE*, 98, 1255–1268, with permission from IEEE. Copyright 2010 IEEE; Fan, Z., and Lu, J.G., *Appl. Phys. Lett.*, 86, 123510, Copyright 2005, American Institute of Physics.)

Table A.23 lists the sensing behavior of different nanostructured metal-oxide devices for various gases and vapors. The devices are MOSFET, FETs, FET structures, NW channel FET, SGFET, VSGFET, TFT, hybrid SGFET, diodes, and Schottky diodes. Different parameters were studied to detect gases such as O_2, NO_2, H_2, NH_3, DMMP, C_2H_4, and C_3H_6 at different concentration levels.

Nanoimprint lithography (NIL), a new lithography paradigm, promises to fabricate sub-10 nm structures with high throughput and low-cost active devices. However, NIL has not been used to fabricate semiconductor devices, so the effects of NIL on the device performance, such as the effects of high pressure, have not been examined. Guo et al. [283] reported nanoscale silicon FETs using NIL. Using this technique, nanostructures are created in a resist by deformation of the shape with embossing, rather than by modification of the resist chemical structures with radiation or self-assembly. During NIL, a mold with nanometer scale features is first pressed into a thin layer of specially prepared polymer, such as polymethyl methacrylate (PMMA) that is coated on the substrate. This process creates a thickness contrast pattern in the polymer. After the mold is removed, an anisotropic etching process is used to transfer the pattern through the entire resist layer by removing the residual polymer in the compressed areas, completing the imprint lithography. The first nanotransistor was fabricated by using this technique.

The integration of a mobile communication device, such as a cellular phone or wireless personal digital assistant, with environmental monitoring sensors to detect humidity, temperature, and UV radiation, has attracted much attention. However, the integration of these devices with an ozone (O_3) gas sensor has not been reported up to date, although ozone is a harmful gas that is produced by UV irradiation as part of sunlight as well as by past-generation photocopiers and laser printers. Wang et al. [226] reported an integration of In_2O_3 nanoparticle–based ozone sensors, operating at room temperature, with GaInN/GaN LED devices. In this case, conventional thermal reactivation of the sensing layer, to be provided by a built-in heating system, was replaced by external UV light sources. This approach paved the way to integrated gas sensitive In_2O_3 films with UV LEDs, resulting in a reduced power consumption, smaller size, and lower production costs, accompanied by a significant miniaturization. A precondition for an integration of metal-oxide films with LEDs is the evaluation of an optimized combination of film thickness and morphology of the gas-sensing layer as well as wavelength and intensity of the light used for photoreduction.

In recent years, there are reports on room-temperature gas-sensing effects on metal-oxide surfaces that are enhanced by UV light and that seems to exhibit some selectivity toward NO_2. This has been demonstrated recently on room-temperature gas sensitivity to short-chain alcohols and to acetone. However, the actual mechanisms that underlie this low-temperature gas-sensing effect are a point of debate [244]. From the application point-of-view, it is important to note that the dissociative room-temperature effect exhibits a significantly different cross sensitivity behavior than the established surface combustion effect that is operative on heated metal-oxide surfaces.

9

Future Devices and Nanostructured Gas-Sensor Arrays

9.1 Future Devices and Gas-Sensor Arrays

Gas sensors have a wide range of applications including in particular environmental monitoring, process and medical control, and quality analysis. Since individual sensors usually cannot fulfill such complex tasks, new instruments, such as electronic noses, have been designed, which typically use several sensors, all of which operate with one of the various possible signal transduction principles [40]. However, in most applications even such sensor arrays are still insufficient in their performance, if compared with established instruments of analytical chemistry, like gas chromatograph/ mass spectrometer couplings (GC/MS). The main problem results from the fact that the individual sensors usually show drift, are not sensitive enough, and detect only certain classes of molecules.

Nanowires and nanowire electronics have drawn lot of interest in recent years and great promise has followed toward their development [284–286]. Developments in nanowire growth have led to the demonstration of a wide range of nanowire materials with precise control of composition, morphology, and electrical properties, and it is believed that this excellent control together with small channel size could yield device performance exceeding that obtained using standard process techniques [287]. While nanowires provide considerable novelty, and are quite useful for studies of the physics of transport in constrained systems, their applicability to future sensors and integrated sensor circuits remains under investigation and difficult to conclude at this stage.

On the other hand, a new generation of devices, the well-known "electronic nose," is obtained by combining several sensors and several transducers in multielement modular sensor systems. Nanotube molecular wires are one such example and are based on carbon nanotubes [288], FET devices based on these carbon nanotubes [289,290], 3C-SiC nanowire FETs [291], graphene nanoribbon FETs [292], and epitaxial graphene transistors on SiC substrates [293]. The gas concentrations are mapped by the sensor array in the output space.

The extraction of information from the sensors is done by the pattern recognition procedure when classification is requested or by nonlinear regression when quantitative analysis is required. The pattern recognition procedure is structured into three parts: feature extraction, classification, and identification. Indeed, it is found in practice that an array of semiconducting sensors in which each element provides a broadband response, with partially overlapping sensitivities, offset from that of the other elements, can provide selectivity for the gaseous species under study.

Chemical sensing of gases is crucial for a number of environmental applications, and a vast number of sensor materials have been developed till date. Normally, they are used for detecting individual or a group of gases. If more complicated sensing is of interest, such as for recording of fungal volatiles in the food industry, it is common to use sensor arrays. Simple and cheap sensors are required for routine applications. Their operating principle is that adsorption of a species onto the surface of the sensor material produces reaction heat or altered surface properties. These properties are frequently obtained either by optical or electrical ways. Electrical detection has an advantage in its simplicity and in the inherent ability to integrate signals. In order to improve the performance of the devices, it is crucial to increase the surface area of the sensor and/or to enhance the signal of the electrical response of the device. Considering the first item, the use of nanoparticle films is appropriate. Regarding signal enhancement, the choice of a proper sensor material, and the development of new sensing technologies, is of course important.

The development of MEMS-based gas sensors is a rapidly growing area of research driven by the numerous advantages that can be realized with such microdevices, including low-power consumption, low cost due to batch fabrication, improved platform reliability, and improved selectivity through the use of arrays. Arrays of microfabricated sensors are attractive for making hand-held electronic noses for applications such as environmental monitoring, pollution measurement, monitoring food freshness, and medical diagnostics [104]. Each of these applications requires sophisticated sensing systems, electronics for data acquisition, and signal processing for species identification and quantification.

Barrettino et al. [246] presented two mixed-signal monolithic gas-sensor microsystems fabricated using 0.8 μm CMOS technology combined with post-CMOS micromachining to form the microhotplates. These systems were implemented in a multichip approach, that is, the microhotplates were placed on one chip, and the necessary driving and signal conditioning circuitry was placed on a separate application-specific integrated circuit (ASIC).

Heat conduction in dielectric or semiconductor thin films has attracted intensive attention in the last decade. It is now well accepted that when the film thickness becomes comparable to, or smaller than, the phonon mean free path, the size effect and interface effect become significant, and the Boltzmann transport equation should be and can be applied to describe phonon transport [294]. Theoretical calculations show that a significant drop in temperature occurs at the interfaces for micro- and nanocylindrical and spherical media. For cylindrical media, the effective thermal conductivity is determined by both the film thickness and the diameter of the inner cylinder. For spherical media, the effective conductivity is mainly determined by the size of the inner sphere. Heat conduction issues are not readily available in the open literature.

9.2 Summary and Conclusions

The complete list of nanostructured metal oxides and their response to different gaseous species and vapors is shown in Table A.24. The metal oxides include: aluminum, cadmium, cerium, cobalt, copper, gallium, indium, iridium, iron, molybdenum, niobium, nickel, tellurium, tin, titanium, tungsten, vanadium, zinc, zirconium, and mixed oxides. The gaseous species are: acetic acid, acetone, acetylene, ammonia, benzene, butane, carbon dioxide, carbon disulfide, carbon monoxide, chlorine, dimethyl methylphosphonate, ethyl alcohol, ethylene, formaldehyde, helium, hexane, hydrogen, hydrogen sulfide, LPG, methane, methanol, methyl cyanide, nitrogen oxides, oxygen, ozone, petrol (gasoline), petroleum ether, propane, propene, propanol, propylene, sulfur dioxide, sulfur hexafluoride, toluene, and xylene. Details of the nanostructures of various metal oxides are listed in Table A.2. As it is seen, the tin oxide nanostructures along with mixed oxides are dominating at present. Tungsten nanostructures very closely follow these two oxides. The intention of this table is to give a clear view of these nanostructures. Comparing Table A.2 with Table A.24, it is easy to draw and select the nanostructures for the desired application.

In spite of many efforts, surface science of metal oxides has not been developed so much when compared to that of metals and semiconductors. As the surfaces of metal oxides are defective in nature and also majority metal oxides are usually insulators, modern surface analytical tools such as electron spectroscopy are not fully applicable to these materials. Scanning Kelvin Probe microscopy (SKPM) using noncontact atomic force microscopy (NCAFM) has an advantage to observe either atomic level images or local electrostatic property even for the surface of nonconductive materials [295]. As many bulk metal-oxide crystals have characteristics of forming a non-stoichiometric long-range structure, structures using metal-oxide surface are expected to play an important role in the field of nanofabrications and their application for gas sensing. The intense efforts by many active scientific groups are working for the development of these nanometal oxide–based devices and it is expected to dominate in immediate future as a practical device.

Appendix: Tables

TABLE A.1

Environmental Safety Standards for Some Selective VOC Vapors
and Pollutant Gases

Gas/Vapor	NIOSH	OSHA	IDLH	Properties
Benzene (C_6H_6)	0.1 ppm (REL) 1 ppm (STEL)	1 ppm 5 ppm (STEL)	500 ppm	Colorless liquid, easily vaporized, dissolves fats
Ethanol (C_2H_5OH)	1,000 ppm (REL)	1,000 ppm	3,300 ppm	Ordinary alcohol as in wine
Methane (CH_4)	90,000 ppm (9%)	90,000 ppm (9%)	—	Component of natural gas. Gas causing mine explosions
Methanol (CH_3OH)	200 ppm (REL) 250 ppm (STEL)	200 ppm	6,000 ppm	An important organic solvent
Propanol (C_3H_7OH)	200 ppm (REL) 250 ppm (STEL)	200 ppm	800 ppm	Alcohol. Clear colorless liquid
Ammonia (NH_3)	25 ppm (REL) 35 ppm (STEL)	50 ppm (PEL)	300 ppm	A colorless gas having a pungent odor
Carbon dioxide (CO_2)	5,000 ppm (REL) 3% (STEL)	5,000 ppm (PEL)	30,000 ppm (3%)	Nonflammable, colorless, odorless, slightly acid gas
Carbon monoxide (CO)	35 ppm (REL) 200 ppm (ceiling)	50 ppm (PEL)	1,200 ppm	Toxic, flammable, colorless, and odorless gas
Hydrogen sulfide (H_2S)	10 ppm (ceiling) for 10 min	20 ppm (ceiling) 50 ppm (10 min)	100 ppm	Colorless, highly flammable, offensive odor (rotten eggs)
Nitrogen compounds (NO, NO_2, NO_x)	25 ppm (REL)	25 ppm (PEL)	100 ppm	NO, N_2O_3, NO_2 (N_2O_4) are extremely toxic gases
Sulfur dioxide (SO_2)	2 ppm (REL) 5 ppm (STEL)	5 ppm (STEL)	100 ppm	Highly irritating, nonflammable, colorless gas

VOC, Volatile organic compounds; NIOSH, National Institute of Occupational Safety and Health; OSHA, Occupational Safety and Health Administration; IDLH, immediately dangerous to life or health; REL, recommended exposure limit; STEL, short-term exposure limit; PEL, permissible exposure limits.

TABLE A.2

Metal-Oxide Nanostructures and Their Syntheses

Metal Oxides	Shape of Nanostructure	Syntheses Process/Technique Used to Grow Nanostructures	References
Al_2O_3	Solid nanosphere particles	By spray drying of droplets method for solid sphere structures	[54]
Al_2O_3	Nanosphere particles with hollow shapes	By spray precipitation method	[54]
Al_2O_3	Bottle-shaped nanoparticles	Spraying precipitation technique by using aluminum chloride and tributylamine	[54]
Al_2O_3	Ultrathin films	Cathodic vacuum arc deposition technique	[55]
Al_2O_3	Nanowires	Synthesized by chemical etching of anodic alumina membrane	[56]
Al_2O_3	Nanoporous thick films	Anodization of plasma-deposited aluminum in oxalic acid	[57]
Bi_2O_3	Nanowires	Synthesized on Au-coated Si substrates by APCVD approach using $Bi(S_2CNEt_2)_3$ in the presence of O_2	[58]
CdO	Nanoneedles	Grown by CVD technique	[59]
CdO	Nanobelts	Controlled solid–vapor process	[60]
CdO	Nanopowder particles of different shapes and sizes	Thermal evaporation technique under atmospheric pressure at 1000°C	[61]
CeO_2	Nanopowder	Chemical syntheses/thick film paste by screenprinting technique	[3]
CeO_2	Nanosized network	By spinodal phase separation of poly-γ-benzyl-L-glutamate	[62]
CeO_2	Nanosized powder	Prepared by mist pyrolysis	[63]
CeO_2	Nanosized particles	Synthesized by hard-template method	[64]
CoO	Nanoparticles	Laser ablation in water and hexane	[65]
Co_3O_4	Nanopowder	Synthesized by a spray pyrolysis method	[66]
Co_3O_4	Nanoparticles	Laser ablation in water and hexane	[65]
Co_3O_4	Nanocrystals in porous SiO_2 matrix	Sol–gel technique	[67]
CuO	Nanowires	Synthesized by direct heating of metal in air ambient	[68]

TABLE A.2 (continued)

Metal-Oxide Nanostructures and Their Syntheses

Metal Oxides	Shape of Nanostructure	Syntheses Process/Technique Used to Grow Nanostructures	References
CuO	Nanopowder	Ultrasonic spray pyrolysis technique	[66]
CuO	Flower-like 3-D nanostructures	Synthesized on copper surface by a simple solution method	[69]
Cu_2O	Nanoparticles	Formed by a chemical reaction between metal thin films and polyamic acid	[70]
Fe_2O_3	Nanothin film	Metal deposition by magnetron sputtering followed by oxidation in air	[71]
Fe_2O_3	Wide variety of nanoshapes	Deposition by novel low-temperature aqueous chemistry technique	[72]
Fe_2O_3	Flower-like nanostructures	Direct immersion of Fe foil into NaCl solution (with CH_3COOH) at 120°C for 16h	[73]
Fe_2O_3	Nanoparticles	Formed by a chemical reaction between metal thin films and polyamic acid	[70]
Fe_2O_3 (Zn)	Thin-film nanoparticles	Liquid-phase deposition	[74]
Fe_2O_3 (Au)	Thin-film nanoparticles	Liquid-phase deposition	[74]
Fe_2O_3 (Pt/Pd/ RuO_2)	Nanoparticles	Fine ground technique	[71]
Ga_2O_3	Nanoparticles	Solvothermal treatment chemical process	[35]
Ga_2O_3	Nanowires	Physical evaporation from bulk gallium target	[75]
Ga_2O_3	Nanotubes, nanowires, and nanopaintbrushes	Exposing molten gallium to appropriate composition of H_2 and O_2 in gaseous phase reaction	[76]
β-Ga_2O_3	Nanowires	Deposition by CVD technique	[77]
β-Ga_2O_3	Nanowires and nanoribbons	Sublimation of gallium metal under argon environment in presence of water vapor	[78]
β-Ga_2O_3	Nanowires, nanobelts, nanosheets, and nanocolumns	Thermal annealing of compacted GaN	[79]
In_2O_3	Nanocrystals	Wet chemical route	[80]
In_2O_3	Nanocrystalline cubic shapes	Film deposition by high-vacuum thermal evaporation (HVTE)	[28]

(continued)

TABLE A.2 (continued)

Metal-Oxide Nanostructures and Their Syntheses

Metal Oxides	Shape of Nanostructure	Syntheses Process/Technique Used to Grow Nanostructures	References
In_2O_3	Nanocrystalline cubic shapes	Sol–gel (SG) spun-coating techniques	[28]
In_2O_3	Nanowires	High-intensity focused electron beam on In nanowires	[28]
In_2O_3	Nanowires	Laser-ablation technique	[81,82]
In_2O_3	Nanowires	Laser-ablation technique	[83]
In_2O_3	Nanowhiskers and nanowires	Heating metal chunks in argon gas flowing atmosphere (at very high temperature)	[84]
In_2O_3	Nanocubes and nano-octahedrons	Heating indium metal chunks in a flowing argon atmosphere	[84]
In_2O_3	Nanocrystalline particles	Laser-ablation technique	[85]
In_2O_3	Nanowires	Laser-ablation-assisted CVD	[86]
In_2O_3	Nanowires	Hot-wall-assisted CVD	[86]
In_2O_3	Nanowires	Vapor-phase deposition technique	[43]
In_2O_3	Nanowires	Condensation from vapor phase	[87]
In_2O_3	Nanowires, nanobelts	Vapor-phase deposition technique	[88]
In_2O_3	Cubic-like nanoparticles	Solvothermal treatment chemical process	[35]
In_2O_3	Octahedral nanocrystals, nanobelts, nanosheets, and nanowires	Hot-wall chemical vapor deposition (HW CVD) technique	[89]
ITO	Nanothin films	Pulsed laser ablation	[90]
ITO	Nanograined thin films	RF magnetron sputtering technique	[91]
IrO_2	Nanothin films	DC sputtering	[92]
MoO_3	Nanorods	Template-directed hydrothermal process	[93]
MoO_3	Nanotubes	Hot-wire CVD technique	[94]
Nb_2O_5	Nanoparticles	Solvothermal treatment chemical process	[35]
NiO	Nanopowder	Ultrasonic spray pyrolysis technique	[66]
NiO	Nanocrystals in porous SiO_2 matrix	Sol–gel technique	[67]
NiO_x	Ultrathin films	Molecular beam deposition technique	[95]

TABLE A.2 (continued)

Metal-Oxide Nanostructures and Their Syntheses

Metal Oxides	Shape of Nanostructure	Syntheses Process/Technique Used to Grow Nanostructures	References
SnO_2	Ultrathin films	Thermal decomposition of Langmuir–Blodgett film precursors	[96]
SnO_2	Ultrathin films	Low-pressure CVD using tin(iv) nitrate precursor	[97]
SnO_2	Wide variety of nanoshapes	Deposition by novel low-temperature aqueous chemistry technique	[72]
SnO_2	Nanowires	Laser-ablation technique with Nd:YAG laser with Sn target in an ambient of oxygen–argon mixture	[98]
SnO_2	Nanowires	Thermal evaporation	[99]
SnO_2	Nanowires	Vapor–solid growth method	[46]
SnO_2	Nanowires	Synthesized with high-temperature chemical reaction method	[100]
SnO_2	Nanowires	Condensation from vapor phase	[87]
SnO_2	Nanorods	By PECVD technique using dibutyltin diacetate as a precursor and subsequent postplasma treatment	[101]
SnO_2	Nanobelts	By thermal evaporation of oxide powders under controlled conditions without the presence of a catalyst	[19]
SnO_2	Nanobelts	Controlled solid–vapor process	[60]
SnO_2	Nanobelts	Thermal evaporation of SnO powder without any catalyst	[102]
SnO_2	Nanoparticles	Hydrothermal treatment sol solution technique	[103]
SnO_2	Nanoparticles	By MOCVD technique using tetramethyltin and oxygen at 500°C on different metal-seeding regions	[104]
SnO_2	Nanoparticles	Sol–gel dip-coating technique	[105]
SnO_2	Nanoparticles	Solvothermal treatment chemical process	[35]
SnO_2	Nanoparticles	By hydrothermally treating α-stannic acid gel in ammonia solution at 200°C	[106]
SnO_2	Nanosized particles	Prepared from supercritical fluid drying technique	[107]

(continued)

TABLE A.2 (continued)

Metal-Oxide Nanostructures and Their Syntheses

Metal Oxides	Shape of Nanostructure	Syntheses Process/Technique Used to Grow Nanostructures	References
SnO_2	Nanopowder thick films	Screenprinting of paste (based on therpineol)	[108]
SnO_2	Nanocrystalline particles	Laser-ablation technique	[109]
SnO_2	Nanograined thin films	Sol–gel technique	[110]
SnO_2	Nanograined thin films	RF sputtering	[108]
SnO_2	Nanocrystalline porous thin films	Sol–gel technique	[111]
SnO_2	Ultrathin films	Sol–gel technique	[112]
SnO_2	Nanothin films	Rheotaxial growth and thermal oxidation (RGTO) process	[113]
$SnO_{1.8}$:Ag	Nanoparticles	By aerosol technique	[114]
SnO_2 (F)	Nanocrystalline films	Sol–gel process	[115]
SnO_2 (Sb)	Thin films of nanocrystals and nanopores	Sol–gel process	[116]
SnO_2 (Pd)	Nano-scaled thin films	Aerosol technology	[117]
SnO_2 (Pd)	Nanosized particles	Prepared from supercritical fluid drying technique	[107]
SnO_2 (Pd)	Nanosized particles	Modified wet chemical route	[118]
SnO_2 (Pd)	Nanopowder	Low-temperature catalyst adding method	[119]
SnO_2 (Pt)	Nanocrystalline particles	Laser-ablation technique	[85]
SnO_2 (CuO)	Nanopowder	Sol–gel technique	[120]
SnO_2 (CuO)	Nanocrystalline thin films	Sol–gel technique	[121]
SnO_2 (CuO)	Nanoribbons	Direct oxidation of Sn at high temperature and *ex situ* mixing of CuO powders	[122]
SnO_2-CuO	Ultrathin films	RF diode sputtering	[123]
SnO_2 (WO_3)	Ultrathin films	Sol–gel technique	[112]
TeO_2	Nanowires	Thermal evaporation of Te in air	[124]
TiO_2	Thin nanofilms	Films by spin coating/preparation by sol–gel technique	[125]
TiO_2	Nanotube array	Anodizing titanium sheet in a 1:7 CH_3COOH and 0.5% HF electrolyte solution	[126]

TABLE A.2 (continued)

Metal-Oxide Nanostructures and Their Syntheses

Metal Oxides	Shape of Nanostructure	Syntheses Process/Technique Used to Grow Nanostructures	References
TiO_2	Nanotubes	Anodization of titanium foil in diluted HF in H_2O	[30]
TiO_2	Nanotube arrays	Anodization of titanium substrates in HF electrolytes	[127]
TiO_2	Spherical nanoparticles and hollow particles	Spraying precipitation technique by using titanium chloride and triethylamine	[54]
TiO_2	Nanopowders	Citrate–nitrate auto combustion method	[128]
TiO_2	Nanoparticles	Hydrolysis of titanium tetra-isopropoxide	[129]
TiO_2	Nanoparticles	Hydrolysis of aqueous $TiCl_4$	[129]
TiO_2	Ultrathin films	Cathodic vacuum arc deposition technique	[55]
TiO_2	Ultrathin films	Low-pressure CVD using titanium(iv) isopropoxide precursor	[97]
TiO_2	Nanocrystalline thin films	Supersonic cluster beam deposition	[130]
TiO_2	Nanocrystalline thin films	DC magnetron sputtering	[131]
TiO_2	Nanofibers	Prepared by electrospinning method	[132]
TiO_2	Sponge-like structures with nanoscale walls/wires	Oxidation of Ti metal in aqueous 10% H_2O_2 at 80°C on Si–SiO_2 platform	[133]
TiO_2 (Al)	Nanopowders	Citrate–nitrate auto combustion method	[128]
TiO_2 (Nb)	Nanopowder	Sol–gel technique	[134]
$WO_{2.72}$	Nanowires	By chemical process	[135]
$WO_{2.72}$	Nanorod film	Solution drop coating	[136]
$W_{18}O_{49}$	Nanotubes	Infrared irradiation of tungsten foils under vacuum on Ta substrate	[137]
$W_{18}O_{49}$	Hollow fibre	By heating WS_2 in the presence of oxygen on surface etched tungsten foil	[138]
$W_{18}O_{49}$	Nanowire	Synthesized by conventional thermal evaporation on ITO without any catalyst	[139]
WO_3	Nanofilms	RF sputtering (with interruptions)	[140]

(continued)

TABLE A.2 (continued)

Metal-Oxide Nanostructures and Their Syntheses

Metal Oxides	Shape of Nanostructure	Syntheses Process/Technique Used to Grow Nanostructures	References
WO_3	Thin films	Vapor-phase condensation onto Si/Si_3N_4 substrates	[141]
WO_3	Thin films	Vacuum evaporation	[142]
WO_3	Nanoparticles	Deposition by supersaturated vapor nucleation onto carbon-coated copper grid	[49,50]
WO_3	Nanoparticles, nanopowders	Advanced reactive gas evaporation technique	[143]
WO_3	Nanopowders	Catalytically modified chemical route with copper and vanadium additives	[144]
WO_3	Nanocrystalline powders	Sol–gel technique using tungstic acid	[145]
WO_3	Nanocrystallites	Reactive RF magnetron sputtering technique	[146]
WO_3	Nanocrystallites	Reactive thermal evaporation method	[147]
WO_3	Nanocrystallites	Aqueous sols of tungsten oxide using ion-exchange method	[148,149]
WO_3	Nanocrystallites	Sol–gel technique using WCl_6	[150]
WO_3	Nanorods	Hot-wire CVD technique	[94]
WO_3	Nanocuboids	One-step hydrothermal process	[151]
WO_3	Nanosized particles	Synthesized by hard template method	[64]
WO_3	Nanosized particle thin films	RF sputtering	[108]
WO_3	Nanocrystalline particles	Laser-ablation technique	[85]
WO_3	Nanopowder thick films	Screenprinted onto silicon substrates	[108]
WO_3	Hollow nanospheres	By controlled hydrolysis of WCl_6 in a coupling solvent	[152]
WO_3	Mesoporous nanostructures	Mesoporous silica template method	[153]
WO_{3-x}	Nanowire networks	Thermal evaporation of tungsten at higher temperatures in the presence of oxygen	[154]
WO_{2+x}	Nanorod film	Hot-filament reactor under flowing argon atmosphere at various filament temperatures	[155]
WO_x	Nanowires	Prepared by electrochemical etching of W wires and subsequently heating in argon atmosphere at high temperature	[156]

TABLE A.2 (continued)

Metal-Oxide Nanostructures and Their Syntheses

Metal Oxides	Shape of Nanostructure	Syntheses Process/Technique Used to Grow Nanostructures	References
WO_x	Nanowires	Flat tungsten plates were reduced with hydrogen at 700°C and later heated in argon atmosphere	[156]
WO_3 (Cu)	Nanocrystalline powder	Chemical route using tungstic acid and copper acetate	[157]
WO_3 (In)	Nanoparticles	By chemical route	[158]
WO_3 (Pd)	Nanoparticles	Produced by advanced reactive gas deposition	[159]
WO_3 (V)	Nanocrystalline powder	Chemical route using tungstic acid and ammonium metavanadate	[157]
V_2O_5	Nanowires	Sol–gel technique of polycondensation of vanadic acid in water	[160]
ZnO	Wide variety of nanoshapes	Deposition by novel low-temperature aqueous chemistry technique	[72]
ZnO	Ultrathin films	Cathodic vacuum arc deposition technique	[55]
ZnO	Nanothin films	Twin laser—twin target pulsed laser deposition technique	[161]
ZnO	Nanothin films	Sol–gel technique	[162]
ZnO	Nanobelts, nanocrystals, nanocantilevers, and nanorods	Controlled solid–vapor process	[60]
ZnO	Nanorods and nanobelts	Electrochemical route directly on zinc substrate	[163]
ZnO	Nanobelts	By RF sputtering technique by using pressed powder target	[164]
ZnO	Hexagonal nanorods	Direct immersion of zinc foil into NaCl solution (with CH_3COOH) at 120°C for 16 h	[73]
ZnO	Nanorods	Molecular beam epitaxy using metal with O_3/O_2 plasma discharge sources and gold droplets as catalyst	[165]
ZnO	Nanorods	Selective vapor–solid growth using a seed-layer technique	[166]
ZnO	Nanorods	Simple wet chemical route at room temperature	[167]
ZnO	Nanorods	Grown on Si (100) wafer in one-step process by direct heating of zinc powder without any catalyst	[168]

(continued)

TABLE A.2 (continued)

Metal-Oxide Nanostructures and Their Syntheses

Metal Oxides	Shape of Nanostructure	Syntheses Process/Technique Used to Grow Nanostructures	References
ZnO	Flower-like nanosheet arrays	Direct immersion of zinc foil with formamide-induced growth at 120°C	[73]
ZnO	Flower-like nanostructures	Synthesized through wet chemical route	[169]
ZnO	Nanowall networks with a honeycomb-like pattern	Evaporation of ZnO mixed with graphite powder in argon environment	[170]
ZnO	Nanowires	By thermal evaporation of zinc under controlled condition without a metal catalyst	[34]
ZnO	Nanowires	Temperature ramping process by using zinc metal in the presence of oxygen and argon ambient	[171]
ZnO	Nanowires	By thermal evaporation and oxidation of metallic zinc powder without metal catalyst or additives	[172]
ZnO	Nanowires	Modified CVD syntheses approach with vapor trapping method (seeding with gold particles)	[173]
ZnO	Nanowires	Thermal evaporation of zinc powders with oxygen carrier gas ambient	[174]
ZnO	Nanowires	Electrochemical and solution deposition technique	[175]
ZnO	Nanowires	Two-step oxygen injection method without catalysts	[176]
ZnO	Nanowires	Low-temperature thermal oxidation of zinc surfaces directly in air	[177]
ZnO	Nanowires	Simple thermal evaporation technique without vacuum in argon environment	[178]
ZnO	Nanowires	Synthesized by physical evaporation method and vertically aligned on templates	[179]
ZnO	Nanowires	Condensation from vapor phase	[87]
ZnO	Nanowires	Controlled evaporation on ZnO:Ga/glass templates in quartz tube	[180]
ZnO	Nanowires	Combination of carbothermal reduction and gold catalyst-mediated heteroepitaxial growth	[181]

TABLE A.2 (continued)

Metal-Oxide Nanostructures and Their Syntheses

Metal Oxides	Shape of Nanostructure	Syntheses Process/Technique Used to Grow Nanostructures	References
ZnO	Nanofibers	Domestic microwave oven using zinc and steel-wool at 2.45 GHz and 1 kW power	[182]
ZnO	Nanofibers	Thermal chemical reactions and vapor transport deposition method in air	[183]
ZnO	Nanofibers	Synthesized by catalyst-free solution route	[184]
ZnO	Nanorods	Deposited by molecular beam epitaxy (MBE) using zinc and O_3/O_2 plasma discharge technique	[185]
ZnO	Nanorods	Catalysis-driven MBE technique	[186]
ZnO	Nanorods	Solution route employing dodecyl benzene sulfonic acid sodium salt as a modifying agent	[187]
ZnO	Nanorods	By hydrothermal method	[188]
ZnO	Nanotubes	Fabricated within nanochannels of porous anodic alumina templates by template wetting process	[189]
ZnO (Cu)	Nanocrystalline thin films	By co-sputtering using ZnO and copper targets in argon environment	[190]
ZrO_2	Hollow particles	Spraying precipitation technique by using zirconyl chloride and tributylamine	[54]
ZrO_2	Ultrathin films	Cathodic vacuum arc deposition technique	[55]
$BaMnO_3$	Nanorods	By composite hydroxide-mediated method	[191]
$BaTiO_3$	Nanopowder	Deposition by laser-ablation technique	[27]
$BaTiO_3$	Nanoparticles	Solvothermal treatment chemical process	[35]
$BaTiO_3$	Nanofibers	By electrospinning technique	[192]
$(CdO)_x(ZnO)_{1-x}$	Nanostructured thin films	Spray pyrolysis method	[193]
$Ce_{1-x}Zr_xO_2$	Nanopowder	New precipitation method using carbon powder	[194]
$Cr_2O_3-TiO_2$	Nanocrystalline thin films	Pulsed-laser deposition	[195]

(continued)

TABLE A.2 (continued)

Metal-Oxide Nanostructures and Their Syntheses

Metal Oxides	Shape of Nanostructure	Syntheses Process/Technique Used to Grow Nanostructures	References
$Fe_2O_3–In_2O_3$	Nanocomposite mixture	Sol–gel technique	[196]
$Fe_2O_3–In_2O_3$	Nanocomposite mixture	Sol–gel technique	[197]
α-Fe_2O_3 + SnO_2	Nanopowder	Mechanical alloying by high-energy ball milling technique	[198]
α-Fe_2O_3 + ZrO_2	Nanopowder	Mechanical alloying by high-energy ball milling technique	[198]
α-Fe_2O_3 + TiO_2	Nanopowder	Mechanical alloying by high-energy ball milling technique	[198]
$Fe_{3-x}Sn_xO_4$	Nanopowder	Wet chemical syntheses	[199]
In-Sn-O	Nanowires	Thermal evaporation	[99]
$La_{0.59}Ca_{0.41}CoO_3$	Nanotubes	Thermal decomposition of the corresponding precursor gel within alumina membranes	[200]
$NiCo_2O_4$	Nanotubes	By thermal decomposition of nitrate precursors within ordered porous alumina template	[201]
$SrTiO_3$	Nanosized material	Synthesized by using high-energy ball milling technique	[202]
$SrTiO_3$	Nanopowders	Thick film screenprinted onto alumina substrates	[203]
W–Ti–O	Thin nanofilms	Deposition by RF sputtering technique	[204]
$xTiO_2$–$(1-x)$ WO_3	Nanopowders	Mechanical alloying by high-energy ball milling at room temperature	[205]
$ZnSnO_3$	Nanowires	Thermal evaporation	[18,39]
$ZrO_2–Fe_2O_3$	Nanopowder	High-energy ball milling	[206]

TABLE A.3

Aluminum-Oxide Nanomaterials

Aluminum Oxides with Additives (Dopant)	Preparation Technique (Grain Size)	Sensing Gas (Vapor)	Operating Temperature	Range of Detection Limits	Sensing Element Form	Sensor Physical Parameter	Response Time	References
Al_2O_3	Anodization of Al in oxalic acid (hexagonal grains of 80 nm)	Cyclo hexane	Room temperature	—	Thick film on ceramic substrate	Impedance measurements (LCR)	—	[57]
Al_2O_3	Anodization of Al in oxalic acid (hexagonal grains of 80 nm)	Cyclo hexene	Room temperature	—	Thick film on ceramic substrate	Impedance measurements (LCR)	—	[57]
Al_2O_3	Anodization of Al in oxalic acid (hexagonal grains of 80 nm)	Benzene	Room temperature	—	Thick film on ceramic substrate	Impedance measurements (LCR)	—	[57]
Al_2O_3	Anodization of Al in oxalic acid (hexagonal grains of 80 nm)	Toluene	Room temperature	—	Thick film on ceramic substrate	Impedance measurements (LCR)	—	[57]
Al_2O_3	Anodization of Al in oxalic acid (hexagonal grains of 80 nm)	*o*-xylene *m*-xylene *p*-xylene	Room temperature	—	Thick film on ceramic substrate	Impedance measurements (LCR)	—	[57]

TABLE A.4

Cadmium-Oxide Nanomaterials

Cadmium Oxides with Additives (Dopant)	Preparation Technique (Grain Size)	Sensing Gas (Vapor)	Operating Temperature	Range of Detection Limits	Sensing Element Form	Sensor Physical Parameter	Response Time	References
CdO	CVD (nanoneedles 40–100 nm diameter, 2–20 μm long)	NO_2	Room temperature	200 ppm	Nanoneedles dispersed onto Si/SiO_2 substrate	I–V electrical characteristics at a fixed bias	—	[66]
CdO (Pt/Pd/ RuO_2)	Finely ground crystallites (of 100 nm size)	Acetone (CH_3– CO–CH_3)	300°C	Up to 20 ppm	Thick films on alumina substrates	Element electrical resistance	~3 s	[71]
CdO (along with SnO_2 bottom layer)	Thermal evaporation	NO_2 NO_x	250°C–300°C	100 ppm	Thin film sensing elements	Film conductance	—	[24,217]
CdO–SnO_2	Solid state reaction of reagents (size ~32 nm)	C_2H_5OH	160°C–460°C	10–100 ppm	Paste onto alumina ceramic tubes	Electrical conductance	~30 s	[32]

TABLE A.5

Cerium-Oxide Nanomaterials

Cerium Oxides with Additives (Dopant)	Preparation Technique (Grain Size)	Sensing Gas (Vapor)	Operating Temperature	Range of Detection Limits	Sensing Element Form	Sensor Physical Parameter	Response Time	References
CeO_2	Mist pyrolysis (particle size 100 nm)	O_2	615°C–1000°C	10^3–10^5 Pa	Nanopowder paste screenprinted on alumina substrates	Element electrical conductivity	0.25–1 s	[63]
CeO_2 (Pt)	Chemical syntheses (nanosized fine powder)	O_2	615°C–1002°C	10^3–10^5 Pa	Nanosized fine particle films screenprinted on alumina substrates	Element electrical conductivity	5–11 s	[3]
CeO_2 on TiO_2	Chemical syntheses (particle size 13.6–70 nm)	O_2	700°C	—	Nanoparticles onto thin cylinders	Element electrical resistance	—	[219]

TABLE A.6

Cobalt-Oxide Nanomaterials

Cobalt Oxides with Additives (Dopant)	Preparation Technique (Grain Size)	Sensing Gas (Vapor)	Operating Temperature	Range of Detection Limits	Sensing Element Form	Sensor Physical Parameter	Response Time	References
Co_3O_4 (in silica matrix)	Sol–gel technique (nanocomposite films of 200–300 nm thickness)	H_2	50°C–300°C	20–850 ppm	Films on Si/Si_3N_4 substrates	Film resistance by interdigitated structure	~7 min	[67]
Co_3O_4 (in silica matrix)	Sol–gel technique (nanocomposite films of 200–300 nm thickness)	CO	Room temperature to 350°C	10–10,000 ppm	Films on SiO_2 glasses	Optical transmission of the films in the range 380 < λ < 780 nm	~1 min	[67]
Co_3O_4 (in silica matrix)	Sol–gel technique (nanocomposite films of 200–300 nm thickness)	CO	50°C–300°C	10–500 ppm	Films on Si/Si_3N_4 substrates	Film resistance by interdigitated structure	~5 min	[67]

TABLE A.7

Copper-Oxide Nanomaterials

Copper Oxides with Additives (Dopant)	Preparation Technique (Grain Size)	Sensing Gas (Vapor)	Operating Temperature	Range of Detection Limits	Sensing Element Form	Sensor Physical Parameter	Response Time	References
CuO–SnO$_2$	Simultaneous vacuum evaporation (film of 320 nm)	H$_2$S	140°C–200°C	5–150 ppm	Thin films on alumina disks	Film resistance and sensitivity	~1 min	[33]
CuO on SnO$_2$	RF reactive sputtering (clusters of 2.5–20 nm)	H$_2$S	130°C	20 ppm	Thin films on borosilicate glass substrates	Film resistance by interdigitated structures	~14 s	[220]

TABLE A.8

Gallium-Oxide Nanomaterials

Gallium Oxides with Additives (Dopant)	Preparation Technique (Grain Size)	Sensing Gas (Vapor)	Operating Temperature	Range of Detection Limits	Sensing Element Form	Sensor Physical Parameter	Response Time	References
β-Ga$_2$O$_3$	CVD (nanowires of few tens of nm diameter)	O$_2$	Room temperature	22–20,000 Pa	Wire-on-electrodes configuration on silicon substrate	Electrical transport properties under 254 nm ultraviolet radiation	~2 s	[77]
β-Ga$_2$O$_3$	Sublimation of gallium under argon environment (nanowires and nanoribbons, 5–90 nm)	Ethanol	<100°C	1,500–6,000 ppm	Wire-on-electrodes configuration on SiO$_2$ substrate	Current–voltage measurement on SiO$_2$ substrate	2.5 s	[78]

TABLE A.9

Indium-Oxide Nanomaterials

Indium Oxides with Additives (Dopant)	Preparation Technique (Grain Size)	Sensing Gas (Vapor)	Operating Temperature	Range of Detection Limits	Sensing Element Form	Sensor Physical Parameter	Response Time	References
In_2O_3	Sol–gel (hydrolytic)	O_3	150°C–350°C	100–3000 ppb	Thin film on sapphire substrate	Electrical conductivity by interdigitated structure	1–4 min	[17]
In_2O_3	Sol–gel (hydrolytic)	NO_2	150°C–350°C	200–3200 ppb	Thin film on sapphire substrate	Electrical conductivity by interdigitated structure	1–4 min	[17]
In_2O_3	Vacuum thermal evaporation (crystallites of 35 nm size)	NO_2	250°C	0.7–7 ppm	Thin film on sapphire substrate	Electrical conductivity by using interdigital structure	~5 min	[28]
In_2O_3	Sol–gel (crystallites of 19 nm size)	NO_2	250°C	0.7–7 ppm	Thin film on sapphire substrate	Electrical conductivity by using interdigital structure	~5 min	[28]
In_2O_3	Vapor phase process (nanowires/ belts of ~100 nm width)	NO_2	25°C–200°C	1–5 ppm	Thin films on alumina substrates	Electrical resistance	~10 min	[88]

(continued)

TABLE A.9 (continued)

Indium-Oxide Nanomaterials

Indium Oxides with Additives (Dopant)	Preparation Technique (Grain Size)	Sensing Gas (Vapor)	Operating Temperature	Range of Detection Limits	Sensing Element Form	Sensor Physical Parameter	Response Time	References
In_2O_3	Vapor phase process (nanowires/ belts of ~100 nm width)	C_2H_5OH	200°C	15–50 ppm	Thin films on alumina substrates	Electrical resistance	~12 min	[88]
In_2O_3	Wet chemical route (crystals of 200 nm)	C_2H_5OH	300°C	1–100 ppm	—	Electrical resistance	~2 s	[80]
In_2O_3	Ultrathin layers (~7 nm diameter particles)	Ozone	Room temperature	40–720 ppb	Films on sapphire on back side of LP MOCVD GaInN/ GaN-based LEDs	Resistance variation with and without UV radiation exposure	~2 min	[226]
In_2O_3	Chemical process (20 nm size particles)	NO_2	250°C	2–20 ppm	Films dispersed in water onto alumina substrates	Conduction current through interdigitated structure	60 s	[35]

Material	Synthesis technique	Gas	Temperature	Concentration	Device configuration	Measurement method	Response time	Ref.
In_2O_3	Non aqueous sol–gel technique (20–30 nm cubes)	O_2	25°C–350°C	2.5%–20%	Films pasted onto alumina substrates	Film electrical resistance	~12 min	[227]
In_2O_3	Condensation from vapor (nanowires of width ~100 nm)	O_3	400°C	280 ppb	Films on alumina substrates	Variation of current	~20 min	[87]
In_2O_3	Laser-ablation and drop-coating technique (particle size 12 nm)	O_3	60°C–84°C	250 ppb	Thick films on alumina substrate	Electrical resistance variation	25–45 s	[85]
In_2O_3	Laser-ablation and drop-coating technique (particle size 12 nm)	NO_2	65°C–135°C	250 ppb	Thick films on alumina substrate	Electrical resistance variation	—	[85]
In_2O_3 (oxygen vacancies)	Single crystalline nanowires	NH_3	Room temperature	1%	Nanowires on metallized SiO_2/Si substrates	Conductance by using I–V measurements	—	[81]
In_2O_3 (Pt)	Non aqueous sol–gel technique (20–30 nm cubes)	O_2	150°C–250°C	2.5%–20%	Films pasted onto alumina substrates	Film electrical resistance	~5 min	[227]

(continued)

TABLE A.9 (continued)

Indium-Oxide Nanomaterials

Indium Oxides with Additives (Dopant)	Preparation Technique (Grain Size)	Sensing Gas (Vapor)	Operating Temperature	Range of Detection Limits	Sensing Element Form	Sensor Physical Parameter	Response Time	References
In_2O_3 (Fe_2O_3)	Electron beam evaporation (grain size 20–70 nm)	Cl_2	250°C–500°C	0.1–5 ppm	Thick films on alumina substrates	Electrical conductivity by gold comb-type electrodes	~2 min	[4]
In_2O_3–MoO_3	Sol–gel (hydrolytic)	O_3	150°C–350°C	100–3000 ppb	Thin film on sapphire substrate	Electrical conductivity by interdigitated structure	1–4 min	[17]
In_2O_3–MoO_3	Sol–gel (hydrolytic)	NO_2	150°C–350°C	200–3200 ppb	Thin film on sapphire substrate	Electrical conductivity by interdigitated structure	1–4 min	[17]

TABLE A.10

Iridium-Oxide Nanomaterials

Iridium Oxide with Additives (Dopant)	Preparation Technique (Grain Size)	Sensing Gas (Vapor)	Operating Temperature	Range of Detection Limits	Sensing Element Form	Sensor Physical Parameter	Response Time	Reference
IrO_2	DC sputtering (20 nm, 100 nm films)	NO_2	130°C	2–100 ppm	FET structure	Work function variation	14 min	[92]

TABLE A.11

Iron-Oxide Nanomaterials

Iron Oxides with Additives (Dopant)	Preparation Technique (Grain Size)	Sensing Gas (Vapor)	Operating Temperature	Range of Detection Limits	Sensing Element Form	Sensor Physical Parameter	Response Time	References
Fe_2O_3 (Zn/Au)	Liquid-phase deposition (particle size 10–35 nm)	Oxygen	350°C–450°C	10 ppm–1%	Thin films on alumina substrates	Element electrical conductivity by interdigitated structure	~1 min	[74]
Fe_2O_3 (Pt/Pd/ RuO_2)	Finely ground crystallites of 100 nm size	Acetone (CH_3–CO–CH_3)	300°C	0.1–20 ppm	Thin films on alumina substrates	Element electrical conductivity variation	—	[71]
Fe_2O_3 (Al_2O_3, La_2O_3)	Mechanical mixing and pressure binding	H_2 CH_4 C_2H_5OH C_3H_8 i-C_4H_{10} LPG	420°C	0.05–0.5 vol%	Embedded structure of porous material on plastic substrates, heater, and SS mesh cap	Element electrical resistivity	~30 s	[228]

α-Fe$_2$O$_3$-SnO$_2$	High-energy ball milling (10 nm)	Ethanol Oxygen	120°C–360°C	100 ppm	Thick films screenprinted onto ceramic substrates	Electrical conductivity by interdigitated structure	—	[44]
α-Fe$_2$O$_3$-ZrO$_2$	High-energy ball milling (10 nm)	Ethanol Oxygen	120°C–360°C	100 ppm	Thick films screenprinted onto ceramic substrates	Electrical conductivity by interdigitated structure	—	[44]
α-Fe$_2$O$_3$-TiO$_2$	High-energy ball milling (10 nm)	Ethanol Oxygen	120°C–360°C	100 ppm	Thick films screenprinted onto ceramic substrates	Electrical conductivity by interdigitated structure	—	[44]
α-Fe$_2$O$_3$x-SnO$_2$-(1 – x)	High-energy ball milling	Ethanol	170°C–340°C	10–1000 ppm	Thick films on alumina substrate by screen printing	Electrical conductivity by interdigitated structure	—	[206,229]
α-Fe$_2$O$_3$x-SnO$_2$-(1 – x)	Mechanical alloying/ high-energy ball milling (particle size 10 nm)	Ethanol	170°C–340°C	10–1000 ppm	Thick films on alumina substrate by screen printing	Electrical conductivity by interdigitated structure	—	[229]

TABLE A.12
Molybdenum-Oxide Nanomaterials

Molybdenum Oxides with Additives (Dopant)	Preparation Technique (Grain Size)	Sensing Gas (Vapor)	Operating Temperature	Range of Detection Limits	Sensing Element Form	Sensor Physical Parameter	Response Time	References
MoO_3	Template directed hydrothermal syntheses (nanorods of 20–280 nm diameter and 0.35–14 μm long)	NO_2	270°C–305°C	1–10 ppm	Nanorods in thick film shape on alumina substrate	Electrical conductivity by interdigitated structure	~15 min	[93]
MoO_3	Template directed hydrothermal syntheses (nanorods of 20–280 nm diameter and 0.35–14 μm long)	NH_3	370°C–420°C	10–100 ppm	Nanorods in thick film shape on alumina substrate	Electrical conductivity by interdigitated structure	~25 min	[93]
MoO_3–In_2O_3	Sol-gel with hydrolytic technique (15 nm diameter and 50 nm long nanotubes and nanorods)	O_3	150°C–350°C	100–3000 ppb	Thin film on sapphire substrate	Electrical conductivity by interdigitated structure	1–4 min	[17]
MoO_3–In_2O_3	Sol-gel with hydrolytic technique (15 nm diameter and 50 nm long nanotubes and nanorods)	NO_2	150°C–350°C	200–3200 ppb	Thin film on sapphire substrate	Electrical conductivity by interdigitated structure	1–4 min	[17]

TABLE A.13

Nickel-Oxide Nanomaterials

Nickel Oxide with Additives (Dopant)	Preparation Technique (Grain Size)	Sensing Gas (Vapor)	Operating Temperature	Range of Detection Limits	Sensing Element Form	Sensor Physical Parameter	Response Time	References
NiO (in silica matrix)	Sol–gel technique (nanocomposite films of 200–300 nm thickness)	H_2	50°C–300°C	20–850 ppm	Films on Si/Si$_3$N$_4$ substrates	Film resistance by interdigitated structure	~10 min	[67]
NiO (in silica matrix)	Sol–gel technique (nanocomposite films of 200–300 nm thickness)	CO	Room temperature to 350°C	10–10,000 ppm	Films on SiO$_2$ glasses	Optical transmission of the films in the range 380 < λ < 780 nm	~1 min	[67]
NiO (in silica matrix)	Sol–gel technique (nanocomposite films of 200–300 nm thickness)	CO	50°C–300°C	10–500 ppm	Films on Si/Si$_3$N$_4$ substrates	Film resistance by interdigitated structure	~5 min	[67]

TABLE A.14

Niobium-Oxide Nanomaterials

Niobium Oxides with Additives (Dopant)	Preparation Technique (Grain Size)	Sensing Gas (Vapor)	Operating Temperature	Range of Detection Limits	Sensing Element Form	Sensor Physical Parameter	Response Time	Reference
Nb_2O_5 (Pt/Pd/RuO_2)	Finely ground crystallites of 100 nm size	Acetone (CH_3—CO—CH_3)	300°C	0.1–20 ppm	Thin films on alumina substrates	Element electrical conductivity variation	~3s	[71]

TABLE A.15

Tellurium-Oxide Nanomaterials

Tellurium Oxides with Additives (Dopant)	Preparation Technique (Grain Size)	Sensing Gas (Vapor)	Operating Temperature	Range of Detection Limits	Sensing Element Form	Sensor Physical Parameter	Response Time	References
TeO_2	Thermal evaporation of metal (nanowires of 30–200 nm diameter and several μm long)	NO_2	Room temperature	10–100 ppm	Nanowires onto oxidized Si substrate	Resistance variation through interdigitated structure with constant voltage	~3 min	[124]
TeO_2	Thermal evaporation of metal (nanowires of 30–200 nm diameter and several μm long)	NH_3	Room temperature	100–500 ppm	Nanowires onto oxidized Si substrate	Resistance variation through interdigitated structure with constant voltage	~7 min	[124]
TeO_2	Thermal evaporation of metal (nanowires of 30–200 nm diameter and several μm long)	H_2S	Room temperature	10–100 ppm	Nanowires onto oxidized Si substrate	Resistance variation through interdigitated structure with constant voltage	~5 min	[124]

TABLE A.16

Tin-Oxide Nanomaterials

Tin Oxides with Additives (Dopant)	Preparation Technique (Grain Size)	Sensing Gas (Vapor)	Operating Temperature	Range of Detection Limits	Sensing Element Form	Sensor Physical Parameter	Response Time	References
SnO_2	Thin metal film of 22 nm deposition and annealing in O_2 ambient	NO_2	130°C	0–100 ppm and 30%–60% RH	Nanoparticulate thin films on oxidized silicon substrates	Work function measurements by Kelvin probe	~10 min	[241]
SnO_2	Ultrathin films by thermal decomposition of LB film precursors	NH_3	Room temperature	5–30 ppm	Si/SnO$_2$/Au and quartz/SnO$_2$/Au sensing elements	I–V and C–V parameters	10 s	[96]
SnO_2	Ultrathin films by thermal decomposition of LB film precursors	H_2S	Room temperature	0.5 ppm	Si/SnO$_2$/Au and quartz/SnO$_2$/Au sensing elements	I–V and C–V parameters	~10 min	[96]
SnO_2	Reactive sputtering (grain size 3–6.5 nm)	NO_2	100°C–350°C	5–800 ppb	Thin films onto alumina substrates	Element electrical resistance	~30 min	[22]
SnO_2	Vapor–solid growth (wires of 100 nm diameter)	NH_3	60°C–300°C	100 ppb–100 ppm	Suspended wire on platinum FIB patches on micro hotplate	Electrical conductivity	10–200 s	[46]
SnO_2	Vapor–solid growth (wires of 100 nm diameter)	CO	60°C–300°C	25–100 ppm	Suspended wire on platinum FIB patches on micro hotplate	Electrical conductivity	—	[46]

SnO$_2$	Vapor–solid growth (wires of 100 nm diameter)	NH$_3$	60°C–300°C	25–100 ppm	Chemiresistor configuration on MEMS-based micro hotplate	Electrical conductivity	15–200 s	[46]
SnO$_2$	Controlled thermal evaporation (nanobelts of 200 nm wide, 20–40 nm thick, few mm long)	CO	400°C	250–500 ppm	Single crystal nanobelts deposited on alumina substrates	Current flow through interdigitated structure biased with constant voltage	~3 min	[19]
SnO$_2$	Controlled thermal evaporation (nanobelts of 200 nm wide, 20–40 nm thick, few mm long)	NO$_2$	200°C	0.5 ppm	Single crystal nanobelts deposited on alumina substrates	Current flow through interdigitated structure biased with constant voltage	~1 min	[19]
SnO$_2$	Controlled thermal evaporation (nanobelts of 200 nm wide, 20–40 nm thick, few mm long)	C$_2$H$_5$OH	400°C	250 ppm	Single crystal nanobelts deposited on alumina substrates	Current flow through interdigitated structure biased with constant voltage	~1 min	[19]

(continued)

TABLE A.16 (continued)

Tin-Oxide Nanomaterials

Tin Oxides with Additives (Dopant)	Preparation Technique (Grain Size)	Sensing Gas (Vapor)	Operating Temperature	Range of Detection Limits	Sensing Element Form	Sensor Physical Parameter	Response Time	References
SnO$_2$	Hydrothermal treatment technique (6–10 nm)	CO H$_2$	150°C–500°C	800 ppm	Paste applied on alumina tube and thin films on alumina substrates	Electrical resistance variations	20–120 s	[103]
SnO$_2$	Chemical process (2–2.5 nm size particles)	CH$_4$	550°C	100–2000 ppm	Films dispersed in water onto alumina substrates	Conduction current through interdigitated structure	~5 min	[35]
SnO$_2$	Chemical process (2–2.5 nm size particles)	CO	400°C	10–100 ppm	Films dispersed in water onto alumina substrates	Conduction current through interdigitated structure	~5 min	[35]
SnO$_2$	Sol–gel technique (nanoparticles of 7–15 nm diameter and pores with 1.6–9 nm)	CO	200°C–400°C	10–100 ppm	Thin films onto alumina substrates	Electrical conductivity	2–5 min	[111]

SnO_2	Sol–gel technique (4.5–9 nm crystallites)	Ethanol	425°C	700 ppm	Nanograined thin films on alumina substrates	Electrical properties	~4 s	[110]
SnO_2	Laser-ablation technique (nanoparticle size 8 nm)	NO_2	70°C–250°C	20–100 ppb	Thin films by drop-coating technique onto alumina substrates	Electrical resistance variations by interdigitated structure	~1 min	[109]
SnO_2	Hydrothermal route (4–15 nm diameter and 100–200 nm long rods)	Ethanol	300°C	10–300 ppm	Nanorod-sensing films	Electrical resistivity	~1 s	[236,237]
SnO_2	Controlled solid–vapor process (rectangular cross section)	CO	200°C 300°C	200 ppm 500 ppm	Nanobelts	Electrical conductivity	~2 min	[60]
SnO_2	Controlled solid–vapor process (rectangular cross section)	C_2H_5OH	200°C 300°C	200 ppm 500 ppm	Nanobelts	Electrical conductivity	~6 min	[60]
SnO_2	LPCVD with tin(iv) nitrate precursor (ultrathin films)	CH_3OH	250°C	6–96 μmol/mol	Si MEMS platforms	Electrical output voltage	~1 min	[97]

(continued)

TABLE A.16 (continued)

Tin-Oxide Nanomaterials

Tin Oxides with Additives (Dopant)	Preparation Technique (Grain Size)	Sensing Gas (Vapor)	Operating Temperature	Range of Detection Limits	Sensing Element Form	Sensor Physical Parameter	Response Time	References
SnO_2	Controlled solid–vapor process (rectangular cross section)	NO_2	200°C 300°C	0.5 ppm	Nanobelts	Electrical conductivity	~8 min	[60]
SnO_2	High-temperature thermal evaporation (rectangular cross section)	NO_2	200°C	0.2–10 ppm	Nanobelts	Electrical conductivity	3 min	[242,243]
SnO_2	High-temperature thermal evaporation (rectangular cross section)	Dimethyl methyl-phosphonate	500°C	53 ppm 78 ppm	Nanobelts	Electrical conductivity	15 s	[242,243]
SnO_2	PECVD with postplasma treatment (rods of 7 nm diameter and 100 nm length)	CO	250°C	50–2000 ppm	Nanorods on to Si–SiO_2 substrate	Electrical conductivity with interdigitated structure	~2 min	[101]
SnO_2	LPCVD with tin(iv) nitrate precursor (ultrathin films)	H_2	250°C	6–96 μmol/mol	Si MEMS platforms	Electrical output voltage	~2 min	[97]

Material	Synthesis	Analyte	Temperature	Concentration	Form	Measurement	Response time	Ref
SnO_2	Thermal evaporation at very high temperature (330 nm width, 80 nm thick, 20 μm long)	H_2	25°C–80°C	2%	Nanorods on to $Si-SiO_2$ substrate	Electrical conductivity with RuO_2 ohmic contacts	~4 min	[102]
SnO_2	Hydrothermally treated particles (grain size 6 nm)	CO H_2	350°C	800 ppm	Thin films on alumina substrates	Electrical resistance variation	—	[106]
SnO_2	LPCVD with tin(iv) nitrate precursor (ultrathin films)	CO	250°C	6–96 μmol/mol	Si MEMS platforms	Electrical output voltage	~1 min	[97]
SnO_2	Condensation from vapor (nanowires of width ~100 nm)	O_3	400°C	280 ppb	Films on alumina substrates	Variation of current	~20 min	[87]
SnO_2	Rheotaxial growth and thermal oxidation (thickness 80 nm)	NO_2	25°C–290°C	510–2120 ppb	Films on $Si-Si_3N_4$ membranes	Electrical resistance variation	45 s	[113]
SnO_2	Screenprinted thick films (nanopowders of 30–50 nm grain size)	C_2H_5OH	110°C–480°C	25–182 ppm	Films on $Si-Si_3N_4$ membranes	Electrical resistance variation	—	[108]

(continued)

TABLE A.16 (continued)
Tin-Oxide Nanomaterials

Tin Oxides with Additives (Dopant)	Preparation Technique (Grain Size)	Sensing Gas (Vapor)	Operating Temperature	Range of Detection Limits	Sensing Element Form	Sensor Physical Parameter	Response Time	References
SnO_2	RF-sputtered thin films (nanoparticles of 15–30 nm grain size)	C_2H_5OH	110°C–480°C	25–182 ppm	Films on Si–Si_3N_4 membranes	Electrical resistance variation	—	[108]
SnO_2	Screenprinted thick films (nanopowders of 30–50 nm grain size)	CH_3–CO–CH_3	110°C–480°C	62–275 ppm	Films on Si–Si_3N_4 membranes	Electrical resistance variation	—	[108]
SnO_2	RF-sputtered thin films (nanoparticles of 15–30 nm grain size)	CH_3–CO–CH_3	110°C–480°C	62–275 ppm	Films on Si–Si_3N_4 membranes	Electrical resistance variation	—	[108]
SnO_2	Screenprinted thick films (nanopowders of 30–50 nm grain size)	NH_3	110°C–480°C	625–1875 ppm	Films on Si–Si_3N_4 membranes	Electrical resistance variation	—	[108]
SnO_2	RF-sputtered thin films (nanoparticles of 15–30 nm grain size)	NH_3	110°C–480°C	625–1875 ppm	Films on Si–Si_3N_4 membranes	Electrical resistance variation	—	[108]

Material	Preparation	Gas	Temperature	Concentration	Application	Measured parameter	Response time	Ref.
SnO_2	Screenprinted thick films (nanopowders of 30–50 nm grain size)	NO_2	110°C–480°C	1–10 ppm	Films on Si-Si_3N_4 membranes	Electrical resistance variation	—	[108]
SnO_2	RF-sputtered thin films (nanoparticles of 15–30 nm grain size)	NO_2	110°C–480°C	1–10 ppm	Films on Si-Si_3N_4 membranes	Electrical resistance variation	—	[108]
SnO_2	Screenprinted thick films (nanopowders of 30–50 nm grain size)	CO	110°C–480°C	1000–3500 ppm	Films on Si-Si_3N_4 membranes	Electrical resistance variation	—	[108]
SnO_2	RF-sputtered thin films (nanoparticles of 15–30 nm grain size)	CO	110°C–480°C	1000–3500 ppm	Films on Si-Si_3N_4 membranes	Electrical resistance variation	—	[108]
SnO_2 (with gold surface layer)	Thin layers by sputter deposition and annealing (thickness 300nm)	H_2 CO NH_3 NO_x O_3 C_2H_2	Room temperature to 410°C	1% 500 ppm 5000 ppm 50 ppm 5000 ppm	Thin-film resistors	Effects of UV illumination on film surface electrical conductivity	~8 min	[244]
SnO_2 (with gold surface layer)	Thin layers by sputter deposition and annealing (thickness 300nm)	NO_2	Room temperature	50 ppm	Thin-film resistors	Effects of UV illumination on film surface electrical conductivity	~2 min	[244]

(continued)

TABLE A.16 (continued)

Tin-Oxide Nanomaterials

Tin Oxides with Additives (Dopant)	Preparation Technique (Grain Size)	Sensing Gas (Vapor)	Operating Temperature	Range of Detection Limits	Sensing Element Form	Sensor Physical Parameter	Response Time	References
SnO_2 with CdO top layer	Thermal evaporation	NO_2 NO_x	250°C–300°C	100 ppm	Thin-film sensing elements	Film conductance	—	[24,217]
$SnO_{1.8}$ (Ag)	Aerosol route (particles in the range of 5–20 nm)	C_2H_5OH	400°C	100 ppb	Thin-film sensing elements	Film electrical conductance	~2 s	[114]
SnO_2 (F)	Sol–gel method (sponge topology with pores of 12–15 nm size)	H_2	20°C	100 ppm	Si-based MEMS structure with micro-bead configuration	Electrical resistance variations	—	[115]
SnO_2 (In)	Thermal evaporation (wires of 70–150 nm diameter and several tens μm long)	C_2H_5OH	400°C	10–1000 ppm	Nanowire paste on to ceramic tubes	Film electrical conductance	2–30 s	[99]
SnO_2 (Ru/Rh)	Sol–gel method (size around 15 nm)	C_4H_{10}	220°C–475°C	2000 ppm	RQ-2 type gas sensor detection meter	Electrical resistance variations	—	[31]
SnO_2 (Ru/Rh)	Sol–gel method (size around 15 nm)	C_2H_5OH	220°C–475°C	100 ppm	RQ-2 type gas sensor detection meter	Electrical resistance variations	—	[31]

Material	Synthesis method	Gas	Temperature	Concentration	Configuration	Detection principle	Response time	Reference
SnO$_2$ (Ru/Rh)	Sol–gel method (size around 15 nm)	Petrol	220°C–475°C	100 ppm	RQ-2 type gas sensor detection meter	Electrical resistance variations	—	[31]
SnO$_2$ (Ru/Rh)	Sol–gel method (size around 15 nm)	CH$_4$	220°C–475°C	5000 ppm	RQ-2 type gas sensor detection meter	Electrical resistance variations	—	[31]
SnO$_2$ (Ru/Rh)	Sol–gel method (size around 15 nm)	H$_2$	220°C–475°C	2000 ppm	RQ-2 type gas sensor detection meter	Electrical resistance variations	—	[31]
SnO$_2$ (Ru/Rh)	Sol–gel method (size around 15 nm)	CO	220°C–475°C	2000 ppm	RQ-2 type gas sensor detection meter	Electrical resistance variations	—	[31]
SnO$_2$ (Pt)	Knudsen evaporation	CO	200°C–450°C	50 ppm	Nanocrystalline thin films on sapphire substrates	Schottky diode structure impedance	—	[21]
SnO$_2$ (Pt)	Laser-ablation and drop-coating technique (particle size 15 nm)	O$_3$	120°C	25 ppb	Thick films on alumina substrate	Electrical resistance variations	40–60 s	[119]
SnO$_2$–Pt	Direct oxidation of Sn–Pt double layer (200 and 3 nm)	C$_2$H$_5$OH	350°C	10–1000 ppm	MEMS-based suspended bridge configuration	Film resistance and sensitivity	—	[245]
SnO$_2$ (Pt/Pd)	Rapid microwave radiation at 2.45 GHz (grain size 53–66 nm)	CO	200°C–600°C	50–1000 ppm	Printed on alumina substrates	Electrical conductivity	—	[207]

(continued)

TABLE A.16 (continued)

Tin–Oxide Nanomaterials

Tin Oxides with Additives (Dopant)	Preparation Technique (Grain Size)	Sensing Gas (Vapor)	Operating Temperature	Range of Detection Limits	Sensing Element Form	Sensor Physical Parameter	Response Time	References
SnO_2 (Pt/Pd)	Rapid microwave radiation at 2.45 GHz (grain size 53–66 nm)	NO_2	300°C–600°C	5 ppm	Printed on alumina substrates	Electrical conductivity	—	[207]
SnO_2 (Pd/Pt)	Sol–gel technique (size 20–110 nm)	CO	230°C	100 ppm	Paste onto cylindrical Al_2O_3 substrates	Electrical resistance	—	[25]
SnO_2 (Pd/Pt)	Sol–gel technique (size 20–110 nm)	NO_2	230°C	2 ppm	Paste onto cylindrical Al_2O_3 substrates	Electrical resistance	—	[25]
SnO_2 (Pd)	Supercritical fluid drying technique (particle size 8–10 nm)	CO	Up to 120°C	300 ppm	Paste onto cylindrical ceramic substrates	Constant current method	—	[107]
SnO_2 (Pd)	Supercritical fluid drying technique (particle size 8–10 nm)	H_2	30°C–120°C	500 ppm	Paste onto cylindrical ceramic substrates	Constant current method	—	[107]
SnO_2 (Pd)	Supercritical fluid drying technique (particle size 8–10 nm)	CH_4	300°C	5000 ppm	Paste onto cylindrical ceramic substrates	Constant current method	—	[107]

Material	Method	Gas	Temperature	Concentration	Substrate	Detection method	Response time	Ref
SnO$_2$ (Pd)	Supercritical fluid drying technique (particle size 8–10 nm)	C$_4$H$_{10}$	120°C–300°C	300 ppm	Paste onto cylindrical ceramic substrates	Constant current method	—	[107]
SnO$_2$ (Pd)	Aerosol technique (thin films of 5–50 nm thickness)	CO	25°C–500°C	0.1%–0.5%	Films on Si and ceramic glass substrates	Electrical resistance variations	1–90 s	[117]
SnO$_2$ (Pd)	Modified wet chemical route (particle size of 8 nm)	CO	200°C	50–700 ppm	Si MEMS-based platform structure	Electrical resistance variations by interdigitated structure	—	[118]
SnO$_2$ (Pd)	Drop-coating method (nanocrystalline film)	CO	290°C–350°C	5 ppm	Thin films on MEMS suspended platform	Electrical resistance variation	<10 s	[36]
SnO$_2$ (Pd)	Low temperature catalyst adding method (nanopowder of 5 nm size)	CH$_4$	400°C	1000–3500 ppm	Thick films on alumina substrates	Electrical resistance variation	—	[119]
SnO$_2$ (Pd)	Drop-coating technique (nanocrystalline film)	CO	275°C	1–5 ppm	MEMS-based micro hotplate platform	Electrical resistance variation	<10 s	[246]
SnO$_2$ (Sm)	Hydrogen reduction method (porous thick films with large surface area)	n-C$_6$H$_{14}$ H$_2$ C$_2$H$_5$OH C$_6$H$_6$ CO	250°C	500 ppm	Thick-film sensing elements	Electrical resistance variations	—	[239]

(continued)

TABLE A.16 (continued)

Tin-Oxide Nanomaterials

Tin Oxides with Additives (Dopant)	Preparation Technique (Grain Size)	Sensing Gas (Vapor)	Operating Temperature	Range of Detection Limits	Sensing Element Form	Sensor Physical Parameter	Response Time	References
SnO$_2$ (Ag/Pt)	Magnetron sputtering (ultrafine particles)	Alcohol	Room temperature	Up to 20% volume concentration	Films on glass substrates	Optical transmittivity	—	[240]
SnO$_2$ (Pt/Pd/ RuO$_2$)	Finely ground crystallites of 100 nm size	Acetone (CH$_3$–CO– CH$_3$)	300°C	0.1–20 ppm	Thin films on alumina substrates	Element electrical conductivity variation	~3 s	[71]
SnO$_2$ (Bi$_2$O$_3$)	High-vacuum evaporation (spongy agglomerate having 5–13 nm crystallites)	H$_2$ CO CH$_4$	450°C	100–5000 ppm	Thin-film sensing elements	Film conductance	10–20 s	[24,247]
SnO$_2$ (In$_2$O$_3$)	Sol–gel dip-coating technique	H$_2$	Room temperature	18–900 ppm	Thin films on MEMS structures	Electrical resistance variations	3–15 min	[105]
SnO$_2$ (In$_2$O$_3$)	RF magnetron sputtering (grain size 10–35 nm)	H$_2$	300°C	1000 ppm	Thin films silk screenprinted onto alumina substrates	Electrical resistance variations	~5 s	[91]
SnO$_2$ (WO$_3$)	Sol–gel spin-coating technique (thin films of 120 nm thick)	NO$_2$	150°C	500 ppm	Ultrathin films on glass substrates	Electrical resistance variations	~2 s	[112]

SnO_2-α-Fe_2O_3	High-energy ball milling (10nm)	Ethanol Oxygen	120°C–360°C	100 ppm	Thick films screenprinted onto ceramic substrates	Electrical conductivity by interdigitated structure	—	[44]
SnO_2-$(1-x)$ α-Fe_2O_3x	Mechanical alloying/high-energy ball milling (particle size 10nm)	Ethanol	170°C–340°C	10–1000 ppm	Thick films on alumina substrate by screen printing	Electrical conductivity by interdigitated structure	—	[229]
SnO_2-CdO	Solid state reaction of reagents (size ~32nm)	C_2H_5OH	160°C–460°C	10–100 ppm	Paste onto alumina ceramic tubes	Electrical conductance	~30 s	[32]
SnO_2 (CuO)	Sol–gel technique (nanopowders of 20nm size)	NO	200°C	250–1000 ppm	Pressed pellets	Electrical resistance variations	~3 min	[120]
SnO_2 (CuO)	Sol–gel technique (nanopowders of 20nm size)	CO_2	400°C	23%–100%	Pressed pellets	Electrical resistance variations	—	[120]
SnO_2 (CuO)	Sol–gel technique (nanopowders of less than 20nm size)	CO_2	450°C	1%–4%	Thin films on quartz substrates	Electrical conductivity by interdigitated structure	~2 min	[248]
SnO_2 (CuO)	Sol–gel technique (nanopowders grain size equal to 20nm)	H_2S	85°C–170°C	100 ppb–10 ppm	Thin films spin coated onto Si_3N_4 bulk micro-machined membranes	Electrical resistance variations	15 s–2 min	[121]

(continued)

TABLE A.16 (continued)

Tin-Oxide Nanomaterials

Tin Oxides with Additives (Dopant)	Preparation Technique (Grain Size)	Sensing Gas (Vapor)	Operating Temperature	Range of Detection Limits	Sensing Element Form	Sensor Physical Parameter	Response Time	References
SnO_2–CuO	Simultaneous vacuum evaporation (film of 320 nm)	H_2S	140°C–200°C	5–150 ppm	Thin films on alumina disks	Film resistance and sensitivity	~1 min	[33]
SnO_2–CuO	Direct oxidation of tin (ribbons with 20–200 nm width and length up to several hundred microns)	H_2S	25°C–70°C	3 ppm	Paste onto alumina tube-like structure	Electrical conductivity by interdigitated structure	15 s	[122]
SnO_2–CuO	Direct oxidation of Sn–Cu double layer (200 and 3 nm)	H_2S	200°C	0.01–3 ppm	MEMS-based suspended bridge configuration	Film resistance and sensitivity	—	[245]
SnO_2–CuO	RF diode sputtering (bilayer structure with dotted top layer)	H_2S	100°C–200°C	20 ppm	Thin films on glass substrates	Electrical resistance variation by interdigitated structure	14 s	[123]

TABLE A.17

Titanium-Oxide Nanomaterials

Titanium Oxides with Additives (Dopant)	Preparation Technique (Grain Size)	Sensing Gas (Vapor)	Operating Temperature	Range of Detection Limits	Sensing Element Form	Sensor Physical Parameter	Response Time	References
TiO_2	Seeded supersonic beam by pulsed micro-plasma cluster source	Ethanol (C_2H_5OH)	170°C–370°C	400–2000 ppm	Nanocrystalline thin films on alumina substrates	Electrical conductivity by interdigitated structure	~3 min	[40]
TiO_2	Supersonic cluster beam deposition (grain size 10 nm)	C_2H_5OH	280°C	300 ppm	Nanocrystalline thin films on alumina substrates	Electrical conductivity by interdigitated structure	60–80 s	[130]
TiO_2	Anodization of titanium foil (46 nm diameter, 400 nm length tubes)	H_2	290°C	1000 ppm	Nanotubes grown from titanium foil	Variation of electrical resistance	~150 s	[30]
TiO_2	LPCVD with tin(iv) nitrate precursor (ultrathin films)	H_2	250°C	6–96 μmol/mol	Si MEMS platforms	Electrical output voltage	~1 min	[97]

(continued)

TABLE A.17 (continued)

Titanium-Oxide Nanomaterials

Titanium Oxides with Additives (Dopant)	Preparation Technique (Grain Size)	Sensing Gas (Vapor)	Operating Temperature	Range of Detection Limits	Sensing Element Form	Sensor Physical Parameter	Response Time	References
TiO_2	Seeded supersonic beam by pulsed micro-plasma cluster source	Methanol (CH_3OH)	260°C–370°C	100–500 ppm	Nanocrystalline thin films on alumina substrates	Electrical conductivity by interdigitated structure	~3 min	[40]
TiO_2	Supersonic cluster beam deposition (grain size 10 nm)	CH_3OH	280°C	800 ppm	Nanocrystalline thin films on alumina substrates	Electrical conductivity by interdigitated structure	130–145 s	[130]
TiO_2	LPCVD with tin(iv) nitrate precursor (ultrathin films)	CH_3OH	250°C	6–96 μmol/mol	Si MEMS platforms	Electrical output voltage	~1 min	[97]
TiO_2	Seeded supersonic beam by pulsed micro-plasma cluster source	Propanol (C_3H_8OH)	200°C–410°C	400–2000 ppm	Nanocrystalline thin films on alumina substrates	Electrical conductivity by interdigitated structure	~3 min	[40]
TiO_2	Supersonic cluster beam deposition (grain size 10 nm)	C_3H_8OH	280°C	800 ppm	Nanocrystalline thin films on alumina substrates	Electrical conductivity by interdigitated structure	130–145 s	[130]

TiO$_2$	Oxidation of Ti metal in aqueous 10% H$_2$O$_2$	O$_2$	200°C	0.3–0.8 mTorr	Sponge-like structure with nanoscale walls/wires	Electrical conductivity	~1 min	[133]
TiO$_2$	Sol–gel (grain size 5–30 nm)	CO	400°C	500 ppm	Screenprinted onto alumina substrates	Variation of electrical resistance	~5 min	[134]
TiO$_2$	LPCVD with tin(iv) nitrate precursor (ultrathin films)	CO	250°C	100–1600 nanomol/mol	Si MEMS platforms	Electrical output voltage	~2 min	[97]
TiO$_2$	DC magnetron sputtering (thin films with nanograins)	NH$_3$	250°C	500 ppm	Thin films onto silicon substrates	Variation of electrical conductance	90 s	[131]
TiO$_2$	Sol–gel (grain size 5–30 nm)	C$_2$H$_5$OH	400°C	100 ppm	Screenprinted onto alumina substrates	Variation of electrical resistance	~2 min	[134]
TiO$_2$ (with Pd layer)	Anodization (tube arrays of 22 nm diameter, 13.5 nm thick, 400 nm length)	H$_2$	24°C	1000 ppm	Nanotubes on titanium sheet	Electrical resistance	~1 min	[125]
TiO$_2$ with CeO$_2$	Chemical syntheses (particle size 13.6–70 nm)	O$_2$	700°C	—	Nanoparticles onto thin cylinders	Element electrical resistance	—	[219]
TiO$_2$ (Al)	Chemical syntheses (particle size of ~100 nm)	CO	600°C	100–500 ppm	Thick films onto alumina substrates	Electrical conductivity by interdigitated structure	~3 min	[128]

(continued)

TABLE A.17 (continued)

Titanium-Oxide Nanomaterials

Titanium Oxides with Additives (Dopant)	Preparation Technique (Grain Size)	Sensing Gas (Vapor)	Operating Temperature	Range of Detection Limits	Sensing Element Form	Sensor Physical Parameter	Response Time	References
TiO$_2$ (Pt)	Sol–gel	Methanol	300°C–500°C	2600–6500 ppm	Nanosized thin films on alumina substrates	Electrical conductivity by interdigitated structure	5.5 min	[29]
TiO$_2$ (Pt)	Sol–gel	Ethanol	300°C–500°C	3000 ppm	Nanosized thin films on alumina substrates	Electrical conductivity by interdigitated structure	5.5 min	[29]
TiO$_2$ (Pt)	Sol–gel	Propanol	300°C–500°C	2100 ppm	Nanosized thin films on alumina substrates	Electrical conductivity by interdigitated structure	5.5 min	[29]
TiO$_2$ (Nb)	Laser-assisted spray pyrolysis (grain size 10 nm)	CO	420°C	20–100 ppm	Screenprinted onto alumina substrates	Variation of electrical conductivity	~3 min	[251]
TiO$_2$ (Nb)	Laser-assisted spray pyrolysis (grain size 10 nm)	NO$_2$	420°C	10 ppm	Screenprinted onto alumina substrates	Variation of electrical conductivity	~3 min	[251]

Material	Preparation	Gas	Temperature	Concentration	Structure	Sensing mechanism	Response time	Reference
TiO_2 (Nb)	Sol–gel (grain size 5–30 nm)	CO	400°C	500 ppm	Screenprinted onto alumina substrates	Variation of electrical resistance	~3 min	[134]
TiO_2 (Nb)	Sol–gel (grain size 5–30 nm)	C_2H_5OH	400°C	100 ppm	Screenprinted onto alumina substrates	Variation of electrical resistance	~3 min	[134]
TiO_2 (Pt/Nb)	Sol–gel (grain size 20–50 nm)	CH_3OH C_2H_5OH	300°C–500°C	500–1250 ppm	Nanocrystalline thin films on alumina and bulk Si substrates	Electrical conductivity by interdigitated structure	20–300 s	[125]
$TiO_2–WO_3$	RF reactive sputtering (nanocrystallites of 30 nm)	NO_2	350°C–800°C	1–20 ppm	Nanoclustered thin films onto Si/alumina substrates	Electrical conductivity by interdigitated structure	—	[254]
$TiO_2–\alpha\text{-}Fe_2O_3$	High-energy ball milling (10 nm)	Ethanol Oxygen	120°C–360°C	100 ppm	Thick films screenprinted onto ceramic substrates	Electrical conductivity by interdigitated structure	—	[44]

TABLE A.18

Tungsten-Oxide Nanomaterials

Tungsten Oxides with Additives (Dopant)	Preparation Technique (Grain Size)	Sensing Gas (Vapor)	Operating Temperature	Range of Detection Limits	Sensing Element Form	Sensor Physical Parameter	Response Time	References
$WO_{2.72}$	Solvothermal syntheses (diameter 5–15 nm and 60 ± 20 nm long nanowires)	H_2S	250°C	10–1000 ppm	Paste coated onto alumina substrates	Current through comb type electrodes	83 s	[135]
$WO_{2.72}$	Solution drop coating of nanorods	NH_3	20°C–250°C	100 ppm	Si MEMS microstructure	Electrical resistance variation	70 s	[136]
WO_3 and WO_x (2.6 $\geq x \leq$ 2.8)	Molecular beam–assisted film deposition (particle diameter 5–9 nm)	CO	280°C–310°C	1000 ppm	Material deposited on top of interdigital capacitor on Si substrate	Impedance measurements	—	[15]
WO_3 and WO_x (2.6 $\geq x \leq$ 2.8)	Molecular beam–assisted film deposition (particle diameter 5–9 nm)	NO	280°C–310°C	1000 ppm	Material deposited on top of interdigital capacitor on Si substrate	Impedance measurements	—	[15]
WO_{3-x}	Thermal evaporation in oxygen atmosphere (wires of 200 nm network)	NO_2	300°C	50 ppb	Drop coating on to alumina substrates	Electrical conductivity by interdigitated structure	~5 min	[154]

Material	Synthesis	Gas	Temperature	Concentration	Structure	Measurement principle	Response time	Reference
WO_{3-x}	Thermal evaporation in oxygen atmosphere (wires of 200 nm network)	H_2S	300°C	10 ppm	Drop coating on to alumina substrates	Electrical conductivity by interdigitated structure	~10 min	[154]
WO_3	Gas phase evaporation and deposition (particle size ~5 nm diameter)	H_2S	27°C–427°C	5–100 ppm	Thin conducting (gold) layer covered by nanoparticles on Al_2O_3	Resistivity fluctuations of nanoparticle film	~10 min	[49,50]
WO_3	Gas phase evaporation and deposition (particle size ~5 nm diameter)	N_2O	27°C–427°C	5–100 ppm	Thin conducting (gold) layer covered by nanoparticles on Al_2O_3	Resistivity fluctuations of nanoparticle film	—	[49,50]
WO_3	Gas phase evaporation and deposition (particle size ~5 nm diameter)	CO	27°C–427°C	5–100 ppm	Thin conducting (gold) layer covered by nanoparticles on Al_2O_3	Resistivity fluctuations of nanoparticle film	—	[49,50]
WO_3	Chemical syntheses (average diameter of 20 nm particles)	H_2S	40°C–250°C	1–1000 ppm	Paste coated onto alumina substrates	Current through comb type electrodes	132 s	[135]
WO_3	One-step hydrothermal process (single crystal nanocuboids)	C_2H_5OH CH_3OH CH_3COCH_3 HCHO	80°C–400°C	1000 ppm	Sample layers fabricated on ceramic tubes	Electrical resistance	21–23 s	[151,201]

(continued)

TABLE A.18 (continued)

Tungsten-Oxide Nanomaterials

Tungsten Oxides with Additives (Dopant)	Preparation Technique (Grain Size)	Sensing Gas (Vapor)	Operating Temperature	Range of Detection Limits	Sensing Element Form	Sensor Physical Parameter	Response Time	References
WO$_3$	One-step hydrothermal process (single crystal nanocuboids)	Petrol	80°C–400°C	1000 ppm	Sample layers fabricated on ceramic tubes	Electrical resistance	78 s	[151,201]
WO$_3$	RF sputtering with interruptions (films of 70–83 nm thickness)	C$_2$H$_5$OH	250°C–300°C	10–500 ppm	Films on silicon substrates	Electrical resistance	~20 s	[140]
WO$_3$	RF sputtering with interruptions (films of 70–83 nm thickness)	NO$_2$	150°C	1 ppm	Films on silicon substrates	Electrical resistance	~25 s	[140]
WO$_3$	RF sputtering with interruptions (films of 70–83 nm thickness)	NH$_3$	250°C	10 ppm	Films on silicon substrates	Electrical resistance	~5 s	[140]

WO_3	Advanced reactive gas evaporation (particle grain size 10–45 nm)	H_2S	Room temperature	10 ppm	Films on alumina substrates	Electrical conduction measurements	~6 min	[143]
WO_3	Advanced reactive gas evaporation (particle grain size 10–45 nm)	H_2S	Room temperature	10 ppm	Thick films screenprinted onto alumina substrates	Electrical conduction measurements	~3 min	[143]
WO_3	Reactive RF magnetron sputtering (average grain size 75–100 nm)	O_3	290°C	0.03–0.8 ppm	Films on SiO_2/Si substrates	Electrical conduction measurements	~1 min	[146]
WO_3	Reactive thermal evaporation (agglomerates of 500 nm long)	NO_2	100°C	100–800 ppb	Films on SiO_2/Si and alumina substrates	Electrical conductivity by interdigitated structure	~10 min	[147]
WO_3	Screenprinted thick films (nanopowders of 30–50 nm grain size)	C_2H_5OH	110°C–480°C	25–182 ppm	Films on $Si–Si_3N_4$ membranes	Electrical resistance variation	~1 min	[108]
WO_3	RF-sputtered thin films (nanoparticles of 15–30 nm grain size)	C_2H_5OH	110°C–480°C	25–182 ppm	Films on $Si–Si_3N_4$ membranes	Electrical resistance variation	~1 min	[108]

(continued)

TABLE A.18 (continued)

Tungsten-Oxide Nanomaterials

Tungsten Oxides with Additives (Dopant)	Preparation Technique (Grain Size)	Sensing Gas (Vapor)	Operating Temperature	Range of Detection Limits	Sensing Element Form	Sensor Physical Parameter	Response Time	References
WO_3	Screenprinted thick films (nanopowders of 30–50 nm grain size)	$CH_3–CO–CH_3$	110°C–480°C	62–275 ppm	Films on Si–Si_3N_4 membranes	Electrical resistance variation	~1 min	[108]
WO_3	RF-sputtered thin films (nanoparticles of 15–30 nm grain size)	$CH_3–CO–CH_3$	110°C–480°C	62–275 ppm	Films on Si–Si_3N_4 membranes	Electrical resistance variation	~1 min	[108]
WO_3	Screenprinted thick films (nanopowders of 30–50 nm grain size)	NH_3	110°C–480°C	625–1875 ppm	Films on Si–Si_3N_4 membranes	Electrical resistance variation	~1 min	[108]
WO_3	RF-sputtered thin films (nanoparticles of 15–30 nm grain size)	NH_3	110°C–480°C	625–1875 ppm	Films on Si–Si_3N_4 membranes	Electrical resistance variation	~1 min	[108]
WO_3	Screenprinted thick films (nanopowders of 30–50 nm grain size)	NO_2	110°C–480°C	1–10 ppm	Films on Si–Si_3N_4 membranes	Electrical resistance variation	~1 min	[108]

WO_3	RF-sputtered thin films (nanoparticles of 15–30 nm grain size)	NO_2	110°C–480°C	1–10 ppm	Films on Si–Si_3N_4 membranes	Electrical resistance variation	~1 min	[108]
WO_3	Screenprinted thick films (nanopowders of 30–50 nm grain size)	CO	110°C–480°C	1000–3500 ppm	Films on Si–Si_3N_4 membranes	Electrical resistance variation	~1 min	[108]
WO_3	RF-sputtered thin films (nanoparticles of 15–30 nm grain size)	CO	110°C–480°C	1000–3500 ppm	Films on Si–Si_3N_4 membranes	Electrical resistance variation	~1 min	[108]
WO_3	Controlled hydrolysis process (hollow spheres of ~400 nm diameter 30 nm thin shell)	H_2S	400°C	10–100 ppm	Thin films coated onto ceramic tubes	Electrical measurements	~15 s	[152]
WO_3	Controlled hydrolysis process (hollow spheres of ~400 nm diameter 30 nm thin shell)	NH_3	400°C	50–5000 ppm	Thin films coated onto ceramic tubes	Electrical measurements	~1.8 min	[152]

(continued)

TABLE A.18 (continued)

Tungsten-Oxide Nanomaterials

Tungsten Oxides with Additives (Dopant)	Preparation Technique (Grain Size)	Sensing Gas (Vapor)	Operating Temperature	Range of Detection Limits	Sensing Element Form	Sensor Physical Parameter	Response Time	References
WO_3	Controlled hydrolysis process (hollow spheres of ~400 nm diameter 30 nm thin shell)	Alcohol	250°C–400°C	50–5000 ppm	Thin films coated onto ceramic tubes	Electrical measurements	~2 min	[152]
WO_3	Controlled hydrolysis process (hollow spheres of ~400 nm diameter 30 nm thin shell)	Acetone	250°C–400°C	50–5000 ppm	Thin films coated onto ceramic tubes	Electrical measurements	~1.3 min	[152]
WO_3	Controlled hydrolysis process (hollow spheres of ~400 nm diameter 30 nm thin shell)	Carbon disulfide (CS_2)	400°C	50–5000 ppm	Thin films coated onto ceramic tubes	Electrical measurements	~1 min	[152]

WO_3	Controlled hydrolysis process (hollow spheres of ~400 nm diameter 30 nm thin shell)	Benzene (C_6H_6)	400°C	500–5000 ppm	Thin films coated onto ceramic tubes	Electrical measurements	~2 min	[152]
WO_3	Controlled hydrolysis process (hollow spheres of ~400 nm diameter 30 nm thin shell)	Acetonitrile (CH_3CN)	400°C	500–5000 ppm	Thin films coated onto ceramic tubes	Electrical measurements	~2.8 min	[152]
WO_3	Controlled hydrolysis process (hollow spheres of ~400 nm diameter 30 nm thin shell)	Petroleum ether	400°C	500–5000 ppm	Thin films coated onto ceramic tubes	Electrical measurements	~1.3 min	[152]
WO_3	Vacuum evaporation (crystallites of 40 nm size)	CH_4	360°C	6%	Vacuum evaporation onto Si/SiO$_2$ substrates	Electrical conductivity micropatterned gold electrodes	—	[142]
WO_3	Sol–gel technique (crystallites of the order of 27–68 nm)	CH_4	200°C–350°C	5000 ppm	Pulverization coating on alumina substrates	Electrical conductivity by interdigitated structure	~1 min	[145]

(continued)

TABLE A.18 (continued)

Tungsten-Oxide Nanomaterials

Tungsten Oxides with Additives (Dopant)	Preparation Technique (Grain Size)	Sensing Gas (Vapor)	Operating Temperature	Range of Detection Limits	Sensing Element Form	Sensor Physical Parameter	Response Time	References
WO_3	Ion-exchange method (nanocrystallite lamella of ~33 nm thickness)	NO_2	200°C–400°C	10–1000 ppb	Slurry screenprinted onto alumina substrates	Electrical resistance variation	15 s–4.3 min	[148,149]
WO_3	Sol–gel technique (crystallites of the order of 20–70 nm)	NO_2	300°C	50–550 ppb	Spin coated onto alumina substrates	Electrical conductivity by interdigitated structure	3 min	[150]
WO_3	Sol–gel technique (crystallites of the order of 27–68 nm)	NO_2	200°C–350°C	5 ppm	Pulverization coating on alumina substrates	Electrical conductivity by interdigitated structure	~1 min	[145]
WO_3	Vapor phase condensation (crystallites of 9–21 nm size)	NO_2	150°C	0.2–0.8 ppm	Vapor phase condensation onto Si/Si_3N_4 substrates	Electrical conductivity by interdigital electrodes	~5 min	[141]
WO_3	Mesoporous silica templates (pore diameter 2–50 nm)	NO_2	100°C–300°C	1 ppm	Printed onto alumina substrates	Electrical resistance variation	~20 min	[153]

Material	Preparation	Gas	Temperature	Concentration	Deposition	Detection method	Response time	Ref.
WO_3	Vacuum evaporation (crystallites of 40 nm size)	NO	250°C	9 ppm	Vacuum evaporation onto Si/SiO_2 substrates	Electrical conductivity micropatterned gold electrodes	~10 min	[142]
WO_3	Laser-ablation and drop-coating technique (particle size 15 nm)	O_3	180°C	250–1000 ppb	Thick films on alumina substrate	Electrical resistance variation	15 s	[85]
WO_3	Sol–gel technique (crystallites of the order of 27–68 nm)	CO	200°C–350°C	500 ppm	Pulverization coating on alumina substrates	Electrical conductivity by interdigitated structure	~1 min	[145]
WO_3 (Cu)	Chemical route (nanocrystalline powder)	NH_3	200°C–300°C	20–200 ppm	Screenprinted onto alumina substrates	Electrical resistance variation	~2 min	[157]
WO_3 (In)	Chemical route (particles of 45 nm)	NO_2	Room temperature to 200°C	1–5 ppm	Screenprinted onto alumina substrates	Electrical conductivity by interdigitated structure	~30 s	[158]
WO_3 (In)	Chemical route (particles of 45 nm)	CO	300°C	100–500 ppm	Screenprinted onto alumina substrates	Electrical conductivity by interdigitated structure	~20 s	[158]

(continued)

TABLE A.18 (continued)

Tungsten-Oxide Nanomaterials

Tungsten Oxides with Additives (Dopant)	Preparation Technique (Grain Size)	Sensing Gas (Vapor)	Operating Temperature	Range of Detection Limits	Sensing Element Form	Sensor Physical Parameter	Response Time	References
WO_3 (In)	Chemical route (particles of 45 nm)	NH_3	300°C	10–500 ppm	Screenprinted onto alumina substrates	Electrical conductivity by interdigitated structure	~60 s	[158]
WO_3 (In)	Chemical route (particles of 45 nm)	C_2H_5OH	300°C	10–500 ppm	Screenprinted onto alumina substrates	Electrical conductivity by interdigitated structure	~3 s	[158]
WO_3 (Pd)	Advanced reactive deposition (particles of ~5 nm)	CO	Room temperature	100 ppm	Thin films onto alumina substrates	Electrical conductivity variations	—	[159]
WO_3 (Pd)	Advanced reactive deposition (particles of ~5 nm)	H_2	Room temperature	500 ppm	Thin films onto alumina substrates	Electrical conductivity variations	—	[159]
WO_3 (Pd)	Advanced reactive deposition (particles of ~5 nm)	SO_2	Room temperature	125 ppm	Thin films onto alumina substrates	Electrical conductivity variations	—	[159]

Material	Method	Gas	Temperature	Concentration	Substrate	Measurement	Time	Ref.
WO$_3$ (Pd)	Advanced reactive deposition (particles of ~5nm)	NO$_2$	Room temperature	100 ppm	Thin films onto alumina substrates	Electrical conductivity variations	—	[159]
WO$_3$ (Pd)	Advanced reactive deposition (particles of ~5nm)	H$_2$S	Room temperature to 200°C	0.5–10 ppm	Thin films onto alumina substrates	Electrical conductivity variations	3.2–10 min	[159]
WO$_3$ (Pd)	Advanced reactive deposition (particles of ~5nm)	HCHO	Room temperature	100 ppm	Thin films onto alumina substrates	Electrical conductivity variations	—	[159]
WO$_3$ (V)	Chemical route (nanocrystalline powder)	NH$_3$	200°C–300°C	20–200 ppm	Screenprinted onto alumina substrates	Electrical resistance variation	~2 min	[157]
W–Ti–O	RF reactive sputtering (crystallites of 50nm)	NO$_2$	350°C–800°C	1–10 ppm	Thin films onto silicon and alumina substrates	Electrical conductivity by interdigitated structure	~2 min	[204]

TABLE A.19

Vanadium-Oxide Nanomaterials

Vanadium Oxides with Additives (Dopant)	Preparation Technique (Grain Size)	Sensing Gas (Vapor)	Operating Temperature	Range of Detection Limits	Sensing Element Form	Sensor Physical Parameter	Response Time	Reference
V_2O_5	Sol–gel	He	Room temperature	0.1–100 mbar	Nanowires on Si–SiO$_2$ substrate on predefined metal connections	Electrical conduction variations $\left(\dfrac{dI}{dt}\right)$	~10 s	[160]

TABLE A.20

Zinc-Oxide Nanomaterials

Zinc Oxides with Additives (Dopant)	Preparation Technique (Grain Size)	Sensing Gas (Vapor)	Operating Temperature	Range of Detection Limits	Sensing Element Form	Sensor Physical Parameter	Response Time	References
ZnO	Chemical precipitation, emulsion, and microemulsion (grain size 20–50 nm)	H_2	300°C–500°C	0.2%	Measured with static state distribution	Electrical resistance	—	[260]
ZnO	Chemical precipitation, emulsion, and microemulsion (grain size 20–50 nm)	C_4H_{10}	300°C–500°C	0.2%	Measured with static state distribution	Electrical resistance	—	[260]
ZnO	Chemical precipitation, emulsion, and microemulsion (grain size 20–50 nm)	Gasoline	300°C–500°C	0.01%	Measured with static state distribution	Electrical resistance	—	[260]
ZnO	Chemical precipitation, emulsion, and microemulsion (grain size 20–50 nm)	C_2H_5OH	300°C–500°C	0.01%	Measured with static state distribution	Electrical resistance	—	[260]

(continued)

TABLE A.20 (continued)

Zinc-Oxide Nanomaterials

Zinc Oxides with Additives (Dopant)	Preparation Technique (Grain Size)	Sensing Gas (Vapor)	Operating Temperature	Range of Detection Limits	Sensing Element Form	Sensor Physical Parameter	Response Time	References
ZnO	Chemical precipitation, emulsion, and microemulsion (grain size 20–50 nm)	SF_6	300°C–500°C	—	Measured with static state distribution	Electrical resistance	—	[260]
ZnO	Controlled thermal evaporation (long nanowires with 25 nm diameter)	Ethanol	300°C	1–200 ppm	Paste onto silicon-based MEMS suspended platform	Electrical resistance by interdigitated structures	~5 s	[34]
ZnO	Temperature ramping process (width 50–200 nm, length 2–5.4 μm wires)	CO	320°C	300–2000 ppm	Nanowires in fingers of the comb-like pattern on SiO_2/Si templates	Electrical response by I–V characteristics	~30 s	[171]
ZnO	Modified CVD approach (60 nm wire diameter and several tens of microns length)	Oxygen	Room temperature to 200°C	0–50 ppm	Field effect transistor	Gate voltage modulation of FET	—	[173]

ZnO	Thermal evaporation of zinc in oxygen environment (40–60 nm wire diameter)	Oxygen	Room temperature	10^{-4}–10^5 Pa	Nanowire transistors	Source-drain current and threshold voltage shifts	—	[174]
ZnO	Controlled evaporation (nanowire diameter 100 nm and 3 μm long)	Oxygen	Room temperature	0.05–500 Torr	Nanowire resistors	Electrical resistance	~12 min	[180]
ZnO	Electrochemical route on zinc (5 nm thick and ~20 nm width)	H_2	Room temperature	250 ppm	Ultrathin arrays	Electrical resistance	~2 min	[163]
ZnO	RF sputtering (nanobelts of 10–100 nm width and several micron long)	Ethanol	200°C–290°C	50–2000 ppm	Copper tube coated with nanobelts	Electrical resistance	~30 s	[164]
ZnO	Wet chemical route (nanorods of 15 nm diameter and 1.0 μm long)	Ethanol	300°C	1–1000 ppm	Films onto ceramic tubes	Electrical resistance	~10 s	[167,169]
ZnO	Wet chemical route (nanorods of 15 nm diameter and 1.0 μm long)	H_2	300°C	50–1000 ppm	Films onto ceramic tubes	Electrical resistance	~10 s	[167,169]

(continued)

TABLE A.20 (continued)

Zinc-Oxide Nanomaterials

Zinc Oxides with Additives (Dopant)	Preparation Technique (Grain Size)	Sensing Gas (Vapor)	Operating Temperature	Range of Detection Limits	Sensing Element Form	Sensor Physical Parameter	Response Time	References
ZnO	Wet chemical route (nanorods of 15 nm diameter and 1.0 μm long)	NH_3	300°C	100–2000 ppm	Films onto ceramic tubes	Electrical resistance	~10 s	[167,169]
ZnO	Chemical route (nanorods of 95 nm diameter)	C_6H_6	100°C–200°C	10 ppb–10 ppm	Paste onto Al_2O_3 tubes	Electrical resistance	~10 s	[187]
ZnO	Chemical route (nanorods of 95 nm diameter)	C_2H_5OH	150°C–450°C	1–500 ppm	Paste onto Al_2O_3 tubes	Electrical resistance	~10 s	[187]
ZnO	Chemical route (nanorods of 95 nm diameter)	$C_6H_5CH_3$	100°C–250°C	1000 ppm	Paste onto Al_2O_3 tubes	Electrical resistance	~10 s	[187]
ZnO	Chemical route (nanorods of 95 nm diameter)	Acetone	150°C–400°C	1000 ppm	Paste onto Al_2O_3 tubes	Electrical resistance	~10 s	[187]

ZnO	Hydrothermal method (nanorods of 60–110 nm diameter and 0.2–1.5 μm long)	H_2S	25°C	0.1–1 ppm	Paste onto Al_2O_3 tubes	Electrical resistance	~20 min	[188]
ZnO	Hydrothermal method (nanorods of 60–110 nm diameter and 0.2–1.5 μm long)	C_2H_5OH	350°C	1–100 ppm	Paste onto Al_2O_3 tubes	Electrical resistance	~2 min	[188]
ZnO	Condensation from vapor (nanowires of width ~100 nm)	O_3	350°C	280 ppb	Films on alumina substrates	Variation of current	~20 min	[87]
ZnO–Au ZnO (Au)	Pulsed laser deposition (thin nanostructures)	Butane	Room temperature	1000 ppm	Thin films on SiO_2 substrates	Optical parameter variations in light signal using silicon photodiode	—	[262]
ZnO–Au	Twin laser-twin target pulsed laser deposition (10–25 nm grain size)	H_2	30°C–90°C	3%	Thin films on Si and SiO_2 substrates	$I–V$ characteristics of Schottky diodes	~10 min	[161]

(continued)

TABLE A.20 (continued)

Zinc-Oxide Nanomaterials

Zinc Oxides with Additives (Dopant)	Preparation Technique (Grain Size)	Sensing Gas (Vapor)	Operating Temperature	Range of Detection Limits	Sensing Element Form	Sensor Physical Parameter	Response Time	References
ZnO–Au	Thermal evaporation of ZnO + graphite (wall thickness 80–140 nm)	H_2	300°C	1000 ppm	Thin film form on GaN/ sapphire substrates	Resistance variation by interdigitated electrodes	~2 min	[170]
ZnO–Pd	Molecular beam epitaxy (30–150 nm diameter and length 2–10 μm)	H_2	Room temperature	10–500 ppm	Multiple nanorods on alumina substrates	I–V characteristics of multiple nanorods	~9 min	[185]
ZnO–Pd	Two step oxygen injection method (50 nm diameter and 3.5 μm long)	Ethanol	230°C	5–500 ppm	Nanowires on patterned ZnO:Ga/ SiO$_2$/Si templates	Electrical resistance	~2 min	[176]
ZnO (Cu)	Co-sputtering process (grain size 5 nm)	CO	350°C	40 ppm	Thin films on glass substrate	Electrical resistance	~30 s	[190]

TABLE A.21

Zirconium-Oxide Nanomaterials

Zirconium Oxides with Additives (Dopant)	Preparation Technique (Grain Size)	Sensing Gas (Vapor)	Operating Temperature	Range of Detection Limits	Sensing Element Form	Sensor Physical Parameter	Response Time	Reference
ZrO_2-α-Fe_2O_3	High-energy ball milling (10 nm)	Ethanol Oxygen	120°C–360°C	100 ppm	Thick films screenprinted onto ceramic substrates	Electrical conductivity by interdigitated structure	—	[44]

TABLE A.22

Mixed-Oxide Nanomaterials

Mixed Oxides with Additives (Dopant)	Preparation Technique (Grain Size)	Sensing Gas (Vapor)	Operating Temperature	Range of Detection Limits	Sensing Element Form	Sensor Physical Parameter	Response Time	References
$BaMnO_3$	Composite hydroxide mediated method (nanorods 150–250 nm long and 20–50 nm width)	O_2	180°C–450°C	200–1000 ppb	Spraying nanorod-ethanol suspension onto Si wafer	Electrical conductance variation	—	[191]
$BaTiO_3$	Stearic acid method (grain size 25 nm)	Humidity (H_2O)	—	30%–90% RH	—	Electrical resistance	—	[265]
$BaTiO_3$ (CuO, La_2O_3, $CaCO_3$)	Laser-ablation technique (grain size less than 50 nm)	CO_2	200°C–600°C	0.5%–7.5%	Thick films screenprinted onto alumina substrates	Electrical conductivity (AC) by interdigitated structure	~3 min	[27]
$CdSnO_3$	Solid state reaction of reagents (size ~32 nm)	C_2H_5OH	160°C–460°C	10–100 ppm	Paste onto alumina ceramic tubes	Electrical conductance	~30 s	[32]
$CdSnO_3$	Solid state reaction of reagents (size ~32 nm)	Gasoline	160°C–460°C	50 ppm	Paste onto alumina ceramic tubes	Electrical conductance	—	[32]

Material	Synthesis	Gas	Temperature	Concentration	Deposition	Measurement	Response time	Ref.
$CdSnO_3$	Solid state reaction of reagents (size ~32 nm)	CO, CH_4, C_4H_{10}, LPG	160°C–460°C	500 ppm	Paste onto alumina ceramic tubes	Electrical conductance	—	[32]
$(CdO)_x(ZnO)_{1-x}$	Spray pyrolysis (thin films of 400–500 nm)	NO_2, CO, CH_4	50°C–350°C	3 ppm, 30 ppm, 3000 ppm	Spray onto alumina substrates	Electrical conductivity by interdigitated structure	~1 min	[193]
$Ce_{1-x}Zr_xO_2$	Chemical route (nanopowder particles of ~20 nm size)	Oxygen	600°C–800°C	2.0×10^4 – 6.0×10^4 Pa	Thick films screenprinted onto ceramic substrates	Electrical conductivity by interdigitated structure	2–200 ms	[194]
$Fe_{3-x}Sn_xO_4$	Wet chemical syntheses (particle size 3–31 nm)	H_2S	150°C	15 ppm	Thick films screenprinted onto alumina substrates	Electrical conductance	~60 min	[199]
$\alpha\text{-}Fe_2O_3\text{-}SnO_2$	High-energy ball milling (10 nm)	Ethanol, Oxygen	120°C–360°C	100 ppm	Thick films screenprinted onto ceramic substrates	Electrical conductivity by interdigitated structure	—	[43]
$\alpha\text{-}Fe_2O_3\text{-}TiO_2$	High-energy ball milling (size 10 nm)	Ethanol, Oxygen	120°C–360°C	100 ppm	Thick films screenprinted onto ceramic substrates	Electrical conductivity by interdigitated structure	—	[43]

(continued)

TABLE A.22 (continued)

Mixed-Oxide Nanomaterials

Mixed Oxides with Additives (Dopant)	Preparation Technique (Grain Size)	Sensing Gas (Vapor)	Operating Temperature	Range of Detection Limits	Sensing Element Form	Sensor Physical Parameter	Response Time	References
$\alpha\text{-Fe}_2\text{O}_3\text{-ZrO}_2$	High-energy ball milling (10 nm)	Ethanol Oxygen	120°C–360°C	100 ppm	Thick films screenprinted onto ceramic substrates	Electrical conductivity by interdigitated structure	—	[43]
$\alpha\text{-Fe}_2\text{O}_3 + \text{SnO}_2$	Mechanical alloying	Ethanol	160°C–360°C	100 ppm	Paste screenprinted onto ceramic substrates	Electrical resistance	—	[198]
$\alpha\text{-Fe}_2\text{O}_3 + \text{ZrO}_2$	Mechanical alloying	Ethanol	160°C–360°C	100 ppm	Paste screenprinted onto ceramic substrates	Electrical resistance	—	[198]
$\alpha\text{-Fe}_2\text{O}_3 + \text{TiO}_2$	Mechanical alloying	Ethanol	160°C–360°C	100 ppm	Paste screenprinted onto ceramic substrates	Electrical resistance	—	[198]
$\text{Fe}_2\text{O}_3\text{-In}_2\text{O}_3$	Sol–gel technique (5–30 nm particles)	CH_3OH	200°C–400°C	100–500 ppm	Deposited onto alumina substrates	Electrical conductivity by interdigitated structure	—	[196]

Material	Technique	Gas	Temperature	Concentration	Substrate	Method	Response time	Ref.
Fe_2O_3–In_2O_3	Sol–gel technique (5–30 nm particles)	C_2H_5OH	200°C–400°C	100–500 ppm	Deposited onto alumina substrates	Electrical conductivity by interdigitated structure	—	[196]
Fe_2O_3–In_2O_3	Sol–gel technique (5–30 nm particles)	CH_4	200°C–400°C	50 ppm	Deposited onto alumina substrates	Electrical conductivity by interdigitated structure	—	[196]
Fe_2O_3–In_2O_3	Sol–gel technique (5–30 nm particles)	CO	250°C–350°C	50 ppm	Deposited onto alumina substrates	Electrical conductivity by interdigitated structure	—	[196]
Fe_2O_3–In_2O_3	Sol–gel technique (5–30 nm particles)	NO_2	100°C–135°C	0.5–5 ppm	Deposited onto alumina substrates	Electrical conductivity by interdigitated structure	—	[196]
Fe_2O_3–In_2O_3	Sol–gel technique (5–30 nm particles)	O_3	100°C–135°C	200 ppm	Deposited onto alumina substrates	Electrical conductivity by interdigitated structure	—	[196]
γ-Fe_2O_3–In_2O_3	Sol–gel technique (thin films of 200 nm thick)	C_2H_5OH	250°C	50 ppm	Spin coated onto alumina substrates	Electrical conductivity by interdigitated structure	~30 min	[197]

(continued)

TABLE A.22 (continued)

Mixed-Oxide Nanomaterials

Mixed Oxides with Additives (Dopant)	Preparation Technique (Grain Size)	Sensing Gas (Vapor)	Operating Temperature	Range of Detection Limits	Sensing Element Form	Sensor Physical Parameter	Response Time	References
γ-Fe_2O_3-In_2O_3	Sol–gel technique (thin films of 200 nm thick)	NO_2	135°C	0.5 ppm	Spin coated onto alumina substrates	Electrical conductivity by interdigitated structure	~30 min	[197]
FeO_x-WO_3-SnO_2	Thick film chemical syntheses	NH_3	330°C	10 ppm	Paste screenprinted onto alumina substrates	Element electrical resistivity	~1 min	[268]
Ga_2O_3-ZnO	Sol–gel technique (thin films of 900 nm thick)	C_3H_6	300°C–600°C	700–1900 ppm	Schottky diodes on SiC substrates	Shift in I–V curve at constant current value	~50 s	[269]
$LaMnO_3$/YSZ/ $Y_{0.17}Tb_{0.17}Zr_{0.66}O_{2-x}$	Electrochemical cell	CO	600°C	29–310 ppm	Solid-state mixed potential sensor	Voltages developed on each electrode	~4 min	[26]
MoO_x-SnO_2	Sol–gel (microcrystals of 10–40 nm)	NO_2	150°C–400°C	10 ppm	Powder pellets and thick films on alumina substrates	Element electrical resistivity	2–10 min	[270]
MoO_x-SnO_2	Sol–gel (micro crystals of 10–40 nm)	O_2	150°C–400°C	20 mbar	Powder pellets and thick films on alumina substrates	Element electrical resistivity	2–10 min	[270]

Material	Synthesis	Gas	Temperature	Concentration	Configuration	Measured parameter	Response time	Ref.
$Ni_{1-x}Co_xMn_xFe_{2-x}O_4$	Chemical syntheses (crystallite size of 10–15 nm)	LPG	100°C–230°C	1000 ppm	Nanoparticle powders applied onto alumina tubes	Element electrical resistivity	<30 s	[16]
$Ni_{1-x}Co_xMn_xFe_{2-x}O_4$	Chemical syntheses (crystallite size of 10–15 nm)	CO	100°C–230°C	—	Nanoparticle powders applied onto alumina tubes	Element electrical resistivity	<30 s	[16]
$Ni_{1-x}Co_xMn_xFe_{2-x}O_4$	Chemical syntheses (crystallite size of 10–15 nm)	Ethanol	100°C–230°C	—	Nanoparticle powders applied onto alumina tubes	Element electrical resistivity	<30 s	[16]
$Ni_{1-x}Co_xMn_xFe_{2-x}O_4$	Chemical syntheses (crystallite size of 10–15 nm)	Methane	100°C–230°C	—	Nanoparticle powders applied onto alumina tubes	Element electrical resistivity	<30 s	[16]
$NiCo_2O_4$ $CuCo_2O_4$ $ZnCo_2O_4$	Porous alumina template process (nanoparticles and hollow tubes of 200 nm diameter 20 nm wall thickness)	Acetic acid SO_2 Ethanol CO Cl_2 NO_2	300°C	400 ppm	Sample layers fabricated on ceramic tubes	Electrical resistance	—	[201]
$Pt/Ce_{0.8}Gd_{0.2}O_{1.9}/Au$	Electrochemical cell	H_2 Propylene CO Propane Methane	600°C	0–500 ppm	Solid-state mixed potential sensor	Voltages developed on each electrode	—	[26]

(continued)

TABLE A.22 (continued)

Mixed-Oxide Nanomaterials

Mixed Oxides with Additives (Dopant)	Preparation Technique (Grain Size)	Sensing Gas (Vapor)	Operating Temperature	Range of Detection Limits	Sensing Element Form	Sensor Physical Parameter	Response Time	References
$SrCe_{0.95}Yb_{0.05}O_3$	Polymer precursor process (grain size 7–70 nm)	H_2	1000°C	3%–100%	Nanostructured thin films on alumina substrate	Electrical conductivity by impedance spectroscopy	60–500 s	[266]
$SrTiO_3$	High-energy ball milling (size 10 nm)	Ethanol Oxygen	120°C–360°C	100 ppm	Thick films screenprinted onto ceramic substrates	Electrical conductivity by interdigitated structure	—	[44]
$SrTiO_3$	High-energy ball milling (grain size 27–75 nm)	Oxygen	40°C	20% (in N_2 ambient)	Thick film paste screenprinted onto alumina substrates	Electrical resistance variation	—	[202]
$SrTiO_3$	High-energy ball milling and high-temperature annealing (grain size 27 nm)	Oxygen	40°C	1%–20%	Thick film paste screenprinted onto alumina substrates	Electrical resistance variation	1.6 min	[203]

Material	Synthesis	Gas	Temperature	Concentration	Structure	Measurement	Response time	Ref.
$SrTiO_3$	High-energy ball milling (grain size ~20 nm)	Oxygen	40°C	1%–20%	Films screenprinted onto alumina substrates	Resistance of the films by using interdigitated electrodes	1.6 min	[267]
$xTiO_2$-$(1-x)WO_3$	High-energy ball milling (nanopowder grain size ~58 nm)	Ethanol	120°C–260°C	100–1000 ppm	Paste screenprinted onto alumina substrates	Resistance of the films by using interdigitated electrodes	15 s	[205]
$ZnSnO_3$	Thermal evaporation (20–90 nm diameter and several μm long)	Ethanol	300°C	1–500 ppm	Nanowires on to ceramic tubular structures	Electrical resistance variation	1 s	[263]
$ZnSnO_3$	Thermal evaporation (~60 nm diameter and several μm long)	Oxygen	—	1.0×10^{-4} – 1.0×10^5 Pa	Nanowires on electrodes separated by a gap ~400 nm on oxidized Si substrate	I–V properties	—	[264]

TABLE A.23

Sensing Behavior of Different Nanostructured Metal-Oxide Devices for Various Gases and Vapors

Nanomaterial/ Structure	Active Device Fabricated	Measuring Gaseous Species	Range of Measurement	Sensor Physical Parameter	Application	References
In$_2$O$_3$ nanowires	FET	O$_2$	—	Current–voltage	Nanoelectronic building blocks, chemical nanosensors	[83]
a-InGaZnO nanothin-film transistors	TFT	—	—	Field effect mobility, threshold voltage shifts	Flat panel displays	[278]
IrO$_2$ nanothin films	Hybrid SGFET	NO$_2$	2–100 ppm	Work function measurements by Kelvin probe	Gas sensing	[92]
SnO$_2$ nanobelt (single)	Diode	—	—	Electrical resistance	Tunable properties using 350 nm UV radiation	[60]
SnO$_2$ nanobelt (single)	FET	O$_2$	—	Source-drain current	Gas sensing	[60]
SnO$_2$ nanobelts	FET structure	DMMP	78 ppb	Electrical conductance	Nerve agent detection	[43,242]
Al–ZnO–Al structure	M-I-M structure (with 30 nm sandwich layer)	—	—	Resistive switching characteristics	Resistance random access memory (RRAM)	[162]
Au–ZnO nanothin films	Schottky diodes	H$_2$	3%	Current–voltage	Gas sensing	[161]
GaN-/ZnO	Diode	—	—	Optical emission	Polariton lasers	[279]
Pt–ZnO nanowires	Schottky diodes	H$_2$	<10,000 ppm	Current–voltage	Gas sensing	[186]
Pt–ZnO nanowires	Schottky diodes	C$_2$H$_4$	<20%	Current–voltage	Tunable properties using 366 nm UV radiation	[186]

Material	Device	Measurement	Concentration	Analyte	Application	Ref.
ZnO nanowires	Photodetector	Photoluminescence spectra	—	—	Photonic, optical and electronic devices	[258]
ZnO nanowire	MOSFET	Drain current–voltage	—	—	Tunable properties using 366 nm UV radiation	[186]
ZnO nanorods	Spin-FET	Source–drain current	—	—	Photo-induced ferromagnets	[186]
ZnO nanowires	VSG-FET	I_D–V_D characteristics including threshold voltage variations	—	—	Tera-level ultra-high-density nanoscale memory and logic devices	[181]
ZnO nanowires	Nanowire channel FET	Gate refresh voltage	1,000–10,000 ppm	NO_2	Gas sensing	[276,277]
ZnO nanowires	Nanowire channel FET	Gate refresh voltage	1,000–10,000 ppm	NH_3	Gas sensing	[276,277]
ZnO–TFT	TFTs with Si_3N_4 and SiO_2 as gate dielectrics	I_D–V_D characteristics, threshold voltage shifts	—	—	Transparent electronics	[280]
$PMMA/ZrO_2$	FET (on silicon substrate)	Current–voltage	—	—	Threshold logic devices	[281]
Pt/Ga_2O_3–ZnO/SiC	Schottky diodes (M-RI-SiC)	Shift in I–V curve at constant current value	700–1,900 ppm	C_3H_6	Gas sensing	[269]
$(ZnO)_x(In_2O_3)_{1-x}$	Thin-film transistors (on silicon substrate)	On/off ratio	—	—	High on/off ratio of the order 10^9	[282]

DMMP, dimethyl methylphosphonate; PMMA, polymethylmethacrylate; SGFET, suspended gate FET; M-RI-SiC, metal-reactive insulator-silicon carbide Schottky diodes; M-I-M, metal-insulator-metal structure; TFT, thin-film transistors; VSG-FET, vertical surround-gate FET.

TABLE A.24

Sensing Behavior of Different Nanostructured Metal Oxides for Various Gases and Vapors

Gases/Vapors	Nano Metal Oxides																			
	Al	Cd	Ce	Co	Cu	Ga	In	Ir	Fe	Mo	Nb	Ni	Sn	Te	Ti	V	W	Zn	Zr	Mx
Acetic acid (CH₃COOH)																				●
Acetone (CH₃COCH₃)		●							●		●		●				●	●		
Acetylene (C₂H₂)													●							
Ammonia (NH₃)							●			●			●	●	●		●	●		
Benzene (C₆H₆)	●												●				●	●		
Butane (*i*-C₄H₁₀)									●											
Butane (*n*-C₄H₁₀)													●					●	●	●
Carbon dioxide (CO₂)													●							●
Carbon disulfide (CS₂)																●				
Carbon monoxide (CO)				●								●	●		●		●	●		●
Chlorine (Cl₂)							●													●
Dimethyl methyl phosphonate (DMMP)													●							

(*continued*)

Ethyl alcohol (C$_2$H$_5$OH)

Formaldehyde (HCHO)

Helium (He)

Hexane (*n*-C$_6$H$_{14}$)

Cyclo Hexane (C$_6$H$_{12}$)

Cyclo Hexene (C$_6$H$_{10}$)

Ethylene (C$_2$H$_4$)

Hydrogen (H$_2$)

Hydrogen sulfide (H$_2$S)

Liquid petroleum gas (LPG)

Methane (CH$_4$)

Methanol (CH$_3$OH)

Methyl cyanide, acetonitrile (CH$_3$CN)

Nitrogen oxides (NO, NO$_2$, NO$_x$)

TABLE A.24 (continued)

Sensing Behavior of Different Nanostructured Metal Oxides for Various Gases and Vapors

Gases/Vapors	Nano Metal Oxides																			
	Al	Cd	Ce	Co	Cu	Ga	In	Ir	Fe	Mo	Nb	Ni	Sn	Te	Ti	V	W	Zn	Zr	Mx
Oxygen (O_2)			●			●	●		●				●		●			●	●	●
Ozone (O_3)							●						●					●		●
Petrol, gasoline (hydrocarbons)										●			●				●	●		●
Petroleum ether ($C_5H_{12} + C_6H_{14}$)																	●			
Propane (C_3H_8)									●				●							●
Propene (C_3H_6)																				●
Propanol (C_3H_7OH)															●					
Propylene (CH_3CHCH_2)																				●
Sulfur dioxide (SO_2)																	●			●
Sulfur hexafluoride (SF_6)																		●		
Toluene ($C_6H_5CH_3$)	●																	●		
omp-Xylene (C_7H_{10})	●																			

References

1. M.-I. Baraton and L. Merhari, Advances in air quality monitoring via nano-technology, *Journal of Nanoparticle Research*, **6**, 107–117, 2004, and the references therein.

2. G. Eranna, B.C. Joshi, D.P. Runthala, and R.P. Gupta, Oxide materials for development of integrated gas sensors—A comprehensive review, *Critical Reviews in Solid State and Materials Sciences*, **29**, 111–188, 2004.

3. S.V. Manorama, N. Izu, W. Shin, I. Matsubara, and N. Murayama, On the platinum sensitization of nanosized cerium dioxide oxygen sensors, *Sensors and Actuators*, **B89**, 299–304, 2003, and the references therein.

4. J. Tamaki, C. Naruo, Y. Yamamoto, and M. Matsuoka, Sensing properties to dilute chlorine gas of indium oxide based thin film sensors prepared by electron beam evaporation, *Sensors and Actuators*, **B83**, 190–194, 2002.

5. R.S. Gohlke, Time-of-flight mass spectrometry and gas-liquid partition chromatography, *Analytical Chemistry*, **31**, 535–541, 1959.

6. R.S. Gohlke and F.W. McLafferty, Early gas chromatography/mass spectrometry, *Journal of the American Society for Mass Spectrometry*, **4**, 367–371, 1993.

7. D.K. Schroder, *Semiconductor Material and Device Characterization*, IEEE and Wiley-Interscience, Hoboken, NJ, 2006.

8. P.R. Griffiths and J.A. de Haseth, *Fourier Transform Infrared Spectrometry*, Wiley-Interscience, John Wiley & Sons, New York, 2007.

9. J. Chou, *Hazardous Gas Monitors: A Practical Guide to Selection, Operation and Applications*, McGraw-Hill Book Company, New York, 2000.

10. N. Taguchi, Published patent application in Japan, S37-47677, October 1962.

11. A.Y. Kovalgin, J. Holleman, G. Iordache, T. Jenneboer, F. Falke, V. Zieren, and M.J. Goossens, Low-power, antifuse-based silicon chemical sensor on a suspended membrane, *Journal of the Electrochemical Society*, **153**, H181–H188, 2006.

12. W.H. Brattain and J. Bardeen, Surface properties of germanium, *Bell System Technical Journal*, **32**, 1–41, 1953.

13. T. Seiyama, A. Kato, K. Fujiishi, and M. Nagatani, A new detector for gaseous components using semiconductive thin films, *Analytical Chemistry*, **34**, 1502–1503, 1962.

14. D.M. Wilson, S. Hoyt, J. Janata, K. Booksh, and L. Obando, Chemical sensors for portable, handheld field instruments, *IEEE Sensors Journal*, **1**, 256–274, 2001.

15. T.P. Huelser, A. Lorke, P. Ifeacho, H. Wiggers, and C. Schulz, Core and grain boundary sensitivity of tungsten-oxide sensor devices by molecular beam assisted particle deposition, *Journal of Applied Physics*, **102**, 124305, 2007.

16. L. Satyanarayana, K.M. Reddy, and S.V. Manorama, Synthesis of nanocrystalline $Ni_{1-x}Co_xMn_xFe_{2-x}O_4$: A material for liquefied petroleum gas sensing, *Sensors and Actuators*, **B89**, 62–67, 2003.

17. A. Gurlo, N. Bârsan, M. Ivanovskaya, U. Weimar, and W. Göpel, In_2O_3 and MoO_3–In_2O_3 thin film semiconductor sensors: Interaction with NO_2 and O_3, *Sensors and Actuators*, **B47**, 92–99, 1998, and the references therein.

18. J.G. Lu, P. Chang, and Z. Fan, Quasi-one-dimensional metal oxide materials—Synthesis, properties and applications, *Materials Science and Engineering*, **R52**, 49–91, 2006, and the references therein.

19. E. Comini, G. Faglia, G. Sberveglieri, Z. Pan, and Z.L. Wang, Stable and highly sensitive gas sensors based on semiconducting oxide nanobelts, *Applied Physics Letters*, **81**, 1869–1871, 2002.

20. G. Bläser, Th. Rühl, C. Diehl, M. Ulrich, and D. Kohl, Nanostructured semiconductor gas sensors to overcome sensitivity limitations due to percolation effects, *Physica A: Statistical Mechanics and Its Applications*, **266**, 218–223, 1999, and the references therein.

21. K.-D. Schierbaum, Engineering of oxide surfaces and metal/oxide interfaces for chemical sensors: Recent trends, *Sensors and Actuators*, **B24–B25**, 239–247, 1995, and the references therein.

22. J. Santos, P. Serrini, B. O'Beirn, and L. Manes, A thin film SnO_2 gas sensor selective to ultra-low NO_2 concentrations in air, *Sensors and Actuators*, **B43**, 154–160, 1997.

23. P.M. Hall, Resistance calculations for thin film patterns, *Thin Solid Films*, **1**, 277–295, 1968.

24. G. Sberveglieri, Recent developments in semiconducting thin-film gas sensors, *Sensors and Actuators*, **B23**, 103–109, 1995, and the references therein.

25. A. Diéguez, A. Vilà, A. Cabot, A. Romano-Rodríguez, J.R. Morante, J. Kappler, N. Bârsan, U. Weimar, and W. Göpel, Influence on the gas sensor performances of the metal chemical states introduced by impregnation of calcinated SnO_2 sol–gel nanocrystals, *Sensors and Actuators*, **B68**, 94–99, 2000.

26. F.H. Garzon, R. Mukundan, and E.L. Brosha, Solid-state mixed potential gas sensors: Theory, experiments and challenges, *Solid State Ionics*, **136–137**, 633–638, 2000.

27. P. Keller, H. Ferkel, K. Zweiacker, J. Naser, J.-U. Meyer, and W. Riehemann, The application of nanocrystalline $BaTiO_3$-composite films as CO_2-sensing layers, *Sensors and Actuators*, **B57**, 39–46, 1999.

28. C. Cantalini, W. Wlodarski, H.T. Sun, M.Z. Atashbar, M. Passacantando, and S. Santucci, NO_2 response of In_2O_3 thin film gas sensors prepared by sol–gel and vacuum thermal evaporation techniques, *Sensors and Actuators*, **B65**, 101–104, 2000.

29. G. Sberveglieri, E. Comini, G. Faglia, M.Z. Atashbar, and W. Wlodarski, Titanium dioxide thin films prepared for alcohol microsensor applications, *Sensors and Actuators*, **B66**, 139–141, 2000.

30. O.K. Varghese, D. Gong, M. Paulose, K.G. Ong, and C.A. Grimes, Hydrogen sensing using titania nanotubes, *Sensors and Actuators*, **B93**, 338–344, 2003.

31. Q. Pan, J. Xu, X. Dong, and J. Zhang, Gas-sensitive properties of nanometer-sized SnO_2, *Sensors and Actuators*, **B66**, 237–239, 2000.

32. Y.-L. Liu, Y. Xing, H.-F. Yang, Z.-M. Liu, Y. Yang, G.-L. Shen, and R.-Q. Yu, Ethanol gas sensing properties of nano-crystalline cadmium stannate thick films doped with Pt, *Analytica Chimica Acta*, **527**, 21–26, 2004.

33. A. Khanna, R. Kumar, and S.S. Bhatti, CuO-doped SnO_2 thin films as hydrogen sulfide gas sensor, *Applied Physics Letters*, **82**, 4388–4390, 2003.

34. Q. Wan, Q.H. Li, Y.J. Chen, T.H. Wang, X.L. He, J.P. Li, and C.L. Lin, Fabrication and ethanol sensing characteristics of ZnO nanowire gas sensors, *Applied Physics Letters*, **84**, 3654–3656, 2004.

35. M. Niederberger, G. Garnweitner, N. Pinna, and G. Neri, Non-aqueous routes to crystalline metal oxide nanoparticles: Formation mechanisms and applications, *Progress in Solid State Chemistry*, **33**, 59–70, 2005, and the references therein.

36. M. Graf, D. Barrettino, M. Zimmermann, A. Hierlemann, H. Baltes, S. Hahn, N. Bârsan, and U. Weimar, CMOS monolithic metal-oxide sensor system comprising a microhotplate and associated circuitry, *IEEE Sensors Journal*, **4**, 9–16, 2004.

37. M.Z. Atashbar, D. Banerji, and S. Singamaneni, Room-temperature hydrogen sensor based on palladium nanowires, *IEEE Sensors Journal*, **5**, 792–797, 2005.

38. W. Zhou, Z. Shao, and W. Jin, Synthesis of nanocrystalline conducting composite oxides based on a non-ion selective combined complexing process for functional applications, *Journals of Alloys and Compounds*, **426**, 368–374, 2006.

39. H. Ogawa, M. Nishikawa, and A. Abe, Hall measurement studies and an electrical conduction model of tin oxide ultrafine particle films, *Journal of Applied Physics*, **53**, 4448–4455, 1982.

40. A.M. Taurino, S. Capone, P. Siciliano, T. Toccoli, A. Boschetti, L. Guerini, and S. Iannotta, Nanostructured TiO_2 thin films prepared by supersonic beams and their application in a sensor array for the discrimination of VOC, *Sensors and Actuators*, **B92**, 292–302, 2003, and the references therein.

41. N. Yamazoe and K. Shimanoe, Theory of power laws for semiconductor gas sensors, *Sensors and Actuators*, **B128**, 566–573, 2008, and the references therein.

42. M.J. Madou and S.R. Morrison, *Chemical Sensing with Solid State Devices*, Academic Press, London, U.K., 1989.

43. E. Comini, Metal oxide nano-crystals for gas sensing, *Analytica Chimica Acta*, **568**, 28–40, 2006, and the references therein.

44. O.K. Tan, W. Cao, Y. Hu, and W. Zhu, Nano-structured oxide semiconductor materials for gas-sensing applications, *Ceramic International*, **30**, 1127–1133, 2004, and the references therein.

45. B.J. Kooi and J.Th.M.D. Hosson, In-situ TEM analysis of the reduction of nanometre-sized Mn_3O_4 precipitates in a metal matrix, *Acta Materialia*, **49**, 765–774, 2001.

46. D.C. Meier, S. Semancik, B. Button, E. Strelcov, and A. Kolmakov, Coupling nanowire chemiresistors with MEMS microhotplate gas sensing platforms, *Applied Physics Letters*, **91**, 063118, 2007.

47. A. Rothschild and Y. Komem, The effect of grain size on the sensitivity of nanocrystalline metal-oxide gas sensors, *Journal of Applied Physics*, **95**, 6374–6380, 2004.

48. C. Xu, J. Tamaki, N. Miura, and N. Yamazoe, Grain size effects on gas sensitivity of porous SnO_2-based elements, *Sensors and Actuators*, **B3**, 147–155, 1991.

49. A. Hoel, L.F. Reyes, P. Heszler, V. Lantto, and C.G. Granqvist, Nanomaterials for environmental applications: Novel WO_3-based gas sensors made by advanced gas deposition, *Current Applied Physics*, **4**, 547–553, 2004.

50. A. Hoel, J. Ederth, J. Kopniczky, P. Heszler, L.B. Kish, E. Olsson, and C.G. Granqvist, Conduction invasion noise in nanoparticle WO_3/Au thin-film devices for gas sensing application, *Smart Materials and Structures*, **11**, 640–644, 2002.

51. J. Riu, A. Maroto, and F.X. Rius, Nanosensors in environmental analysis, *Talanta*, **69**, 288–301, 2006, and the references therein.

52. S. Iizuka, S. Ooka, A. Nakata, M. Mizuhata, and S. Deki, Development of fabrication process for metal oxide with nano-structure by the liquid-phase infiltration (LPI) method, *Electrochimica Acta*, **51**, 802–808, 2005.

53. T. Maegawa, T. Yamauchi, T. Hara, H. Tsuchiya, and M. Ogawa, Strain effects on electronic bandstructures in nanoscaled silicon: From bulk to nanowire, *IEEE Transactions on Electron Devices*, **56**, 553–559, 2009.

54. T.-C. Chou, T.-R. Ling, M.-C. Yang, and C.-C. Liu, Micro and nano scale metal oxide hollow particles produced by spray precipitation in a liquid–liquid system, *Materials Science and Engineering*, **A359**, 24–30, 2003.

55. B.K. Tay, Z.W. Zhao, and D.H.C. Chua, Review of metal oxide films deposited by filtered cathodic vacuum arc technique, *Materials Science and Engineering*, **R52**, 1–48, 2006.

56. J.K. Han, J. Kim, Y.C. Choi, K.-S. Chang, J. Lee, H.J. Youn, and S.D. Bu, Structure of alumina nanowires synthesized by chemical etching of anodic alumina membrane, *Physica E: Low-Dimensional Systems and Nanostructures*, **36**, 140–146, 2007.

57. M. Kocanda, M.J. Haji-Sheikh, and D.S. Ballantine, Detection of cyclic volatile organic compounds using single-step anodized nanoporous alumina sensors, *IEEE Sensors Journal*, **9**, 836–841, 2009.

58. X.-P. Shen, S.-K. Wu, H. Zhao, and Q. Liu, Synthesis of single-crystalline Bi_2O_3 nanowires by atmospheric pressure chemical vapor deposition approach, *Physica E: Low-Dimensional Systems and Nanostructures*, **39**, 133–136, 2007.

59. X. Liu, C. Li, S. Han, J. Han, and C. Zhou, Synthesis and electronic transport studies of CdO nanoneedles, *Applied Physics Letters*, **82**, 1950–1952, 2003.

60. Z.L. Wang, Functional oxide nanobelts: Materials, properties and potential applications in nanosystems and biotechnology, *Annual Review of Physical Chemistry*, **55**, 159–196, 2004, and the references therein.

61. A.K. Srivastava, S. Pandey, K.N. Sood, S.K. Halder, and R. Kishore, Novel growth morphologies of nano- and micro-structured cadmium oxide, *Materials Letters*, **62**, 727–730, 2008.

62. C. Sanchez, G.J.A.A. Soler-Illia, F. Ribot, and D. Grosso, Design of functional nano-structured materials through the use of controlled hybrid organic-inorganic interfaces, *Comptes Rendus Chimie*, **6**, 1131–1151, 2003, and the references therein.

63. N. Izu, W. Shin, and N. Murayama, Fast response of resistive-type oxygen gas sensors based on nano-sized ceria powder, *Sensors and Actuators*, **B93**, 449–453, 2003.

64. E. Rossinyol, J. Arbiol, F. Peiró, A. Cornet, J.R. Morante, B. Tian, T. Bo, and D. Zhao, Nanostructured metal oxides synthesized by hard template method for gas sensing applications, *Sensors and Actuators*, **B109**, 57–63, 2005.

65. T. Tsuji, T. Hamagami, T. Kawamura, J. Yamaki, and M. Tsuji, Laser ablation of cobalt and cobalt oxides in liquids: Influence of solvent on composition of prepared nanoparticles, *Applied Surface Science*, **243**, 214–219, 2005.

66. S.W. Oh, H.J. Bang, Y.C. Bae, and Y.-K. Sun, Effect of calcination temperature on morphology, crystallinity and electrochemical properties of nano-crystalline metal oxides (Co_3O_4, CuO, and NiO) prepared via ultrasonic spray pyrolysis, *Journal of Power Sources*, **173**, 502–509, 2007.

67. C. Cantalini, M. Post, D. Buso, M. Guglielmi, and A. Martucci, Gas sensing properties of nanocrystalline NiO and Co_3O_4 in porous silica sol–gel films, *Sensors and Actuators*, **B108**, 184–192, 2005.

68. Y. Xia, P. Yang, Y. Sun, Y. Wu, B. Mayers, B. Gates, Y. Yin, F. Kim, and H. Yan, One-dimensional nanostructures: Synthesis, characterization, and applications, *Advanced Materials*, **15**, 353–389, 2003, and the references therein.

69. X. Zhang, Y.-G. Guo, W.-M. Liu, and J.-C. Hao, CuO three-dimensional flower-like nanostructures: Controlled synthesis and characterization, *Journal of Applied Physics*, **103**, 114304, 2008.

70. J.H. Kim, E.K. Kim, C.H. Lee, M.S. Song, Y.-H. Kim, and J. Kim, Electrical properties of metal-oxide semiconductor nano-particle device, *Physica E: Low-Dimensional Systems and Nanostructures*, **26**, 432–435, 2005.

71. S.V. Ryabtsev, A.V. Shaposhnick, A.N. Lukin, and E.P. Domashevskaya, Application of semiconductor gas sensors for medical diagnostics, *Sensors and Actuators*, **B59**, 26–29, 1999.

72. L. Vayssieres, Advanced semiconductor nanostructures, *Comptes Rendus Chimie*, **9**, 691–701, 2006, and the references therein.

73. C. Yan and D. Xue, Solution growth of nano- to microscopic ZnO on Zn, *Journal of Crystal Growth*, **310**, 1836–1840, 2008.

74. G. Neri, A. Bonavita, S. Galvagno, Y.X. Li, K. Galatsis, and W. Wlodarski, O_2 sensing properties of Zn- and Au-doped Fe_2O_3 thin films, *IEEE Sensors Journal*, **3**, 195–198, 2003.

75. H.Z. Zhang, Y.C. Kong, Y.Z. Wang, X. Du, Z.G. Bai, J.J. Wang, D.P. Yu, Y. Ding, Q.L. Hang, and S.Q. Feng, Ga_2O_3 nanowires prepared by physical evaporation, *Solid State Communications*, **109**, 677–682, 1999.

76. S. Sharma and M.K. Sunkara, Direct synthesis of gallium oxide tubes, nanowires, and nanopaintbrushes, *Journal of American Chemical Society*, **124**, 12288–12293, 2002.

77. P. Feng, X.Y. Xue, Y.G. Liu, Q. Wan, and T.H. Wang, Achieving fast oxygen response in individual β-Ga_2O_3 nanowires by ultraviolet illumination, *Applied Physics Letters*, **89**, 112114, 2006.

78. M.-F. Yu, M.Z. Atashbar, and X. Chen, Mechanical and electrical characterization of β-Ga_2O_3 nanostructures for sensing applications, *IEEE Sensors Journal*, **5**, 20–25, 2005.

79. W.-S. Jung, H.U. Joo, and B.-K. Min, Growth of β-gallium oxide nanostructures by the thermal annealing of compacted gallium nitride powder, *Physica E: Low-Dimensional Systems and Nanostructures*, **36**, 226–230, 2007.

80. P. Feng, X.Y. Xue, Y.G. Liu, and T.H. Wang, Highly sensitive ethanol sensors based on {100}-bounded In_2O_3 nanocrystals due to face contact, *Applied Physics Letters*, **89**, 243514, 2006.

81. D. Zhang, C. Li, X. Liu, S. Han, T. Tang, and C. Zhou, Doping dependent NH_3 sensing of indium oxide nanowires, *Applied Physics Letters*, **83**, 1845–1847, 2003.

82. C. Li, D. Zhang, S. Han, X. Liu, T. Tang, and C. Zhou, Diameter-controlled growth of single-crystalline In_2O_3 nanowires and their electronic properties, *Advanced Materials*, **15**, 143–146, 2003.

83. D. Zhang, C. Li, S. Han, X. Liu, T. Tang, W. Jin, and C. Zhou, Electronic transport studies of single-crystalline In_2O_3 nanowires, *Applied Physics Letters*, **82**, 112–114, 2003.

84. Y. Zhang, H. Ago, J. Liu, M. Yumura, K. Uchida, S. Ohshima, S. Iijima, J. Zhu, and X. Zhang, The synthesis of In, In_2O_3 nanowires and In_2O_3 nanoparticles with shape-controlled, *Journal of Crystal Growth*, **264**, 363–368, 2004, and the references therein.

85. T.K.H. Starke and G.S.V. Coles, High sensitivity ozone sensors for environmental monitoring produced using laser ablated nanocrystalline metal oxides, *IEEE Sensors Journal*, **2**, 14–19, 2002.

86. E. Stern, G. Cheng, S. Guthrie, D. Turner-Evans, E. Broomfield, B. Lei, C. Li, D. Zhang, C. Zhou, and M.A. Reed, Comparison of laser-ablation and hot-wall chemical vapour deposition techniques for nanowire fabrication, *Nanotechnology*, **17**, S246–S252, 2006.

87. G. Sberveglieri, C. Baratto, E. Comini, G. Faglia, M. Ferroni, A. Ponzoni, and A. Vomiero, Synthesis and characterization of semiconducting nanowires for gas sensing, *Sensors and Actuators*, **B121**, 208–213, 2007.

88. S. Bianchi, E. Comini, M. Ferroni, G. Faglia, A. Vomiero, and G. Sberveglieri, Indium oxide quasi-monodimensional low temperature gas sensor, *Sensors and Actuators*, **B118**, 204–207, 2006.

89. G. Cheng, E. Stern, S. Guthrie, M.A. Reed, R. Klie, Y. Hao, G. Meng, and L. Zhang, Indium oxide nanostructures, *Applied Physics A: Materials Science & Processing*, **A85**, 233–240, 2006, and the references therein.

90. A.D. Bonis, A. Galasso, V. Marotta, S. Orlando, A. Santagata, R. Teghil, S. Veronesi, P. Villani, and A. Giardini, Pulsed laser ablation of indium tin oxide in the nano and femtosecond regime: Characterization of transient species, *Applied Surface Science*, **252**, 4632–4636, 2006.

91. K.S. Yoo, S.H. Park, and J.H. Kang, Nano-grained thin-film indium tin oxide gas sensors for H_2 detection, *Sensors and Actuators*, **B108**, 159–164, 2005.

92. A. Karthigeyan, R.P. Gupta, K. Scharnagl, M. Burgmair, M. Zimmer, T. Sulima, S. Venkataraj, S.K. Sharma, and I. Eisele, Iridium oxide as low temperature NO_2-sensitive material for work function-based gas sensors, *IEEE Sensors Journal*, **4**, 189–194, 2004, and the references therein.

93. A.M. Taurino, A. Forleo, L. Francioso, P. Siciliano, M. Stalder, and R. Nesper, Synthesis, electrical characterization, and gas sensing properties of molybdenum oxide nanorods, *Applied Physics Letters*, **88**, 152111, 2006, and the references therein.

94. A.C. Dillon, A.H. Mahan, R. Deshpande, J.L. Alleman, J.L. Blackburn, P.A. Parillia, M.J. Heben, C. Engtrakul, K.E.H. Gilbert, K.M. Jones, R. To, S.-H. Lee, and J.H. Lehman, Hot-wire chemical vapor synthesis for a variety of nanomaterials with novel applications, *Thin Solid Films*, **501**, 216–220, 2006.

95. A. Neubecker, T. Pompl, T. Doll, W. Hansch, and I. Eisele, Ozone-enhanced molecular beam deposition of nickel oxide (NiO) for sensor applications, *Thin Solid Films*, **310**, 19–23, 1997.

96. S. Choudhury, C.A. Betty, K.G. Girija, and S.K. Kulshreshtha, Room temperature gas sensitivity of ultrathin SnO_2 films prepared from Langmuir-Blodgett film precursors, *Applied Physics Letters*, **89**, 071914, 2006.

97. M.Y. Afridi, J.S. Suehle, M.E. Zaghloul, D.W. Berning, A.R. Hefner, R.E. Cavicchi, S. Semancik, C.B. Montgomery, and C.J. Taylor, A monolithic CMOS microhotplate-based gas sensor system, *IEEE Sensors Journal*, **2**, 644–655, 2002.

98. Z. Liu, D. Zhang, S. Han, C. Li, T. Tang, W. Jin, X. Liu, B. Lei, and C. Zhou, Laser ablation synthesis and electron transport studies of tin oxide nanowires, *Advanced Materials*, **15**, 1754–1757, 2003.

99. X.Y. Xue, Y.J. Chen, Y.G. Liu, S.L. Shi, Y.G. Wang, and T.H. Wang, Synthesis and ethanol sensing properties of indium-doped tin oxide nanowires, *Applied Physics Letters*, **88**, 201907, 2006.

100. R. Liu, Y. Chen, F. Wang, L. Cao, A. Pan, G. Yang, T. Wang, and B. Zou, Stimulated emission from trapped excitons in SnO_2 nanowires, *Physica E: Low-Dimensional Systems and Nanostructures*, **39**, 223–229, 2007.

101. H. Huang, O.K. Tan, Y.C. Lee, T.D. Tran, M.S. Tse, and X. Yao, Semiconductor gas sensor based on tin oxide nanorods prepared by plasma-enhanced chemical vapor deposition with postplasma treatment, *Applied Physics Letters*, **87**, 163123, 2005.
102. L.L. Fields, J.P. Zheng, Y. Cheng, and P. Xiong, Room-temperature low-power hydrogen sensor based on a single tin dioxide nanobelt, *Applied Physics Letters*, **88**, 263102, 2006.
103. N.S. Baik, G. Sakai, N. Miura, and N. Yamazoe, Hydrothermally treated sol solution of tin oxide for thin-film gas sensor, *Sensors and Actuators*, **B63**, 74–79, 2000.
104. B. Panchapakesan, D.L. DeVoe, M.R. Widmaier, R. Cavicchi, and S. Semancik, Nanoparticle engineering and control of tin oxide microstructures for chemical microsensor applications, *Nanotechnology*, **12**, 336–349, 2001, and the references therein.
105. S. Shukla, R. Agrawal, H.J. Cho, S. Seal, L. Ludwig, and C. Parish, Effect of ultraviolet radiation exposure on room-temperature hydrogen sensitivity of nanocrystalline doped tin oxide sensor incorporated into microelectromechanical systems device, *Journal of Applied Physics*, **97**, 054307, 2005.
106. G. Sakai, N.S. Baik, N. Miura, and N. Yamazoe, Gas sensing properties of tin oxide thin films fabricated from hydrothermally treated nanoparticles dependence of CO and H_2 response on film thickness, *Sensors and Actuators*, **B77**, 116–121, 2001.
107. F. Lu, Y. Liu, M. Dong, and X. Wang, Nanosized tin oxide as the novel material with simultaneous detection towards CO, H_2 and CH_4, *Sensors and Actuators*, **B66**, 225–227, 2000.
108. P. Ivanov, M. Stankova, E. Llobet, X. Vilanova, J. Brezmes, I. Gràcia, C. Cané, J. Calderer, and X. Correig, Nanoparticle metal-oxide films for micro-hotplate-based gas sensor systems, *IEEE Sensors Journal*, **5**, 798–809, 2005.
109. T.K.H. Starke and G.S.V. Coles, Reduced response times using adsorption kinetics and pulsed-mode operation for the detection of oxides of nitrogen with nanocrystalline SnO_2 sensors, *IEEE Sensors Journal*, **3**, 447–453, 2003.
110. O.K. Varghese and L.K. Malhotra, Electrode-sample capacitance effect on ethanol sensitivity of nano-grained SnO_2 thin films, *Sensors and Actuators*, **B53**, 19–23, 1998.
111. Z. Jin, H.-J. Zhou, Z.-L. Jin, R.F. Savinell, and C.-C. Liu, Application of nanocrystalline porous tin oxide thin film for CO sensing, *Sensors and Actuators*, **B52**, 188–194, 1998.
112. J. Kaur, S.C. Roy, and M.C. Bhatnagar, Highly sensitive SnO_2 thin film NO_2 gas sensor operating at low temperature, *Sensors and Actuators*, **B123**, 1090–1095, 2007.
113. J.G. Partridge, M.R. Field, A.Z. Sadek, K. Kalantar-zadeh, J.D. Plessis, M.B. Taylor, A. Atanacio, K.E. Prince, and D.G. McCulloch, Fabrication, structural characterization and testing of a nanostructured tin oxide gas sensor, *IEEE Sensors Journal*, **9**, 563–568, 2009.
114. R.K. Joshi and F.E. Kruis, Influence of Ag particle size on ethanol sensing of $SnO_{1.8}$:Ag nanoparticle films: A method to develop parts per billion level gas sensors, *Applied Physics Letters*, **89**, 153116, 2006.
115. C.-H. Han, S.-D. Han, I. Singh, and T. Toupance, Micro-bead of nano-crystalline F-doped SnO_2 as a sensitive hydrogen gas sensor, *Sensors and Actuators*, **B109**, 264–269, 2005.

116. D. Zhang, Z. Deng, J. Zhang, and L. Chen, Microstructure and electrical properties of antimony-doped tin oxide thin film deposited by sol–gel process, *Materials Chemistry and Physics*, **98**, 353–357, 2006.

117. G. Korotcenkov, V. Brinzari, J. Schwank, and A. Cerneavschi, Possibilities of aerosol technology for deposition of SnO_2-based films with improved gas sensing characteristics, *Materials Science and Engineering*, **C19**, 73–77, 2002.

118. J.C. Belmonte, J. Manzano, J. Arbiol, A. Cirera, J. Puigcorbé, A. Vilà, N. Sabaté, I. Gràcia, C. Cané, and J.R. Morante, Micromachined twin gas sensor for CO and O_2 quantification based on catalytically modified nano-SnO_2, *Sensors and Actuators*, **B114**, 881–892, 2006.

119. S.-J. Hong and J.-I. Han, Low-temperature catalyst adding for tin-oxide nano-structure gas sensors, *IEEE Sensors Journal*, **5**, 12–19, 2005.

120. G. Zhang and M. Liu, Effect of particle size and dopant on properties of SnO_2-based gas sensors, *Sensors and Actuators*, **B69**, 144–152, 2000.

121. B. Esfandyarpour, S. Mohajerzadeh, A.A. Khodadadi, and M.D. Robertson, Ultra-sensitive tin-oxide microsensors for H_2S detection, *IEEE Sensors Journal*, **4**, 449–454, 2004.

122. X. Kong and Y. Li, High sensitivity of CuO modified SnO_2 nanoribbons to H_2S at room temperature, *Sensors and Actuators*, **B105**, 449–453, 2005.

123. A. Chowdhuri, V. Gupta, and K. Sreenivas, Enhanced catalytic activity of ultra-thin CuO islands on SnO_2 films for fast response H_2S gas sensors, *IEEE Sensors Journal*, **3**, 680–686, 2003.

124. Z. Liu, T. Yamazaki, Y. Shen, T. Kikuta, N. Nakatani, and T. Kawabata, Room temperature gas sensing of p-type TeO_2 nanowires, *Applied Physics Letters*, **90**, 173119, 2007.

125. E. Comini, G. Faglia, G. Sberveglieri, Y.X. Li, W. Wlodarski, and M.K. Ghantasala, Sensitivity enhancement towards ethanol and methanol of TiO_2 films doped with Pt and Nb, *Sensors and Actuators*, **B64**, 169–174, 2000.

126. G.K. Mor, M.A. Carvalho, O.K. Varghese, M.V. Pishko, and C.A. Grimes, A room-temperature TiO_2-nanotube hydrogen sensor able to self-clean photoactively from environmental contamination, *Journal of Materials Research*, **19**, 628–634, 2004.

127. J. Zhao, X. Wang, R. Chen, and L. Li, Fabrication of titanium oxide nanotube arrays by anodic oxidation, *Solid State Communications*, **134**, 705–710, 2005.

128. Y.J. Choi, Z. Seeley, A. Bandyopadhyay, S. Bose, and S.A. Akbar, Aluminum-doped TiO_2 nano-powders for gas sensors, *Sensors and Actuators*, **B124**, 111–117, 2007.

129. S. Nakade, S. Kambe, M. Matsuda, Y. Saito, T. Kitamura, Y. Wada, and S. Yanagida, Electron transport in electrodes consisting of metal oxide nano-particles filled with electrolyte solution, *Physica E: Low-Dimensional Systems and Nanostructures*, **14**, 210–214, 2002.

130. T. Toccoli, S. Capone, L. Guerini, M. Anderle, A. Boschetti, E. Iacob, V. Micheli, P. Siciliano, and S. Iannotta, Growth of titanium dioxide films by cluster supersonic beams for VOC sensing applications, *IEEE Sensors Journal*, **3**, 199–205, 2003.

131. B. Karunagaran, P. Uthirakumar, S.J. Chung, S. Velumani, and E.-K. Suh, TiO_2 thin film gas sensor for monitoring ammonia, *Materials Characterization*, **58**, 680–684, 2007.

132. P. Viswanathamurthi, N. Bhattarai, C.K. Kim, H.Y. Kim, and D.R. Lee, Ruthenium doped TiO_2 fibers by electrospinning, *Inorganic Chemistry Communications*, **7**, 679–682, 2004.

133. A.S. Zuruzi, A. Kolmakov, N.C. MacDonald, and M. Moskovits, Highly sensitive gas sensor based on integrated titania nanosponge arrays, *Applied Physics Letters*, **88**, 102904, 2006.

134. A. Ruiz, A. Calleja, F. Espiell, A. Cornet, and J.R. Morante, Nanosized Nb-TiO$_2$ gas sensors derived from alkoxides hydrolization, *IEEE Sensors Journal*, **3**, 189–194, 2003.

135. C.S. Rout, M. Hegde, and C.N.R. Rao, H$_2$S sensors based on tungsten oxide nanostructures, *Sensors and Actuators*, **B128**, 488–493, 2008, and the references therein.

136. Y.S. Kim, S.-C. Ha, K. Kim, H. Yang, S.-Y. Choi, Y.T. Kim, J.T. Park, C.H. Lee, J. Choi, J. Paek, and K. Lee, Room-temperature semiconductor gas sensor based on nonstoichiometric tungsten oxide nanorod film, *Applied Physics Letters*, **86**, 213105, 2005.

137. Y. Li, Y. Bando, and D. Golberg, Quasi-aligned single-crystalline W$_{18}$O$_{49}$ nanotubes and nanowires, *Advanced Materials*, **15**, 1294–1296, 2003.

138. W.B. Hu, Y.Q. Zhu, W.K. Hsu, B.H. Chang, M. Terrones, N. Grobert, H. Terrones, J.P. Hare, H.W. Kroto, and D.R.M. Walton, Generation of hollow crystalline tungsten oxide fibres, *Applied Physics A: Materials Science & Processing*, **70**, 231–233, 2000, and the references therein.

139. K. Huang, Q. Pan, F. Yang, S. Ni, and D. He, Synthesis and field-emission properties of the tungsten oxide nanowire arrays, *Physica E: Low-Dimensional Systems and Nanostructures*, **39**, 219–222, 2007.

140. V. Khatko, J. Calderer, E. Llobet, and X. Correig, New technology of metal oxide thin film preparation for chemical sensor application, *Sensors and Actuators*, **B109**, 128–134, 2005.

141. C. Cantalini, L. Lozzi, M. Passacantando, and S. Santucci, The comparative effect of two different annealing temperatures and times on the sensitivity and long-term stability of WO$_3$ thin films for detecting NO$_2$, *IEEE Sensors Journal*, **3**, 171–179, 2003.

142. A.H. Jayatissa and L. Mapa, Sensitivity of tungsten oxide thin films for nitric oxide and methane gases, in *Proceedings of the IEEE Conference on Electro/Information Technology*, Windsor, Ontario, Canada, 186–189, 2009.

143. J.L. Solis, S. Saukko, L. Kish, C.G. Granqvist, and V. Lantto, Semiconductor gas sensors based on nanostructured tungsten oxide, *Thin Solid Films*, **391**, 255–260, 2001.

144. I. Jiménez, M.A. Centeno, R. Scotti, F. Morazzoni, A. Cornet, and J.R. Morante, NH$_3$ interaction with catalytically modified nano-WO$_3$ powders for gas sensing applications, *Journal of the Electrochemical Society*, **150**, H72–H80, 2003.

145. I. Jiménez, J. Arbiol, A. Cornet, and J.R. Morante, Structural and gas-sensing properties of WO$_3$ nanocrystalline powders obtained by a sol–gel method from tungstic acid, *IEEE Sensors Journal*, **2**, 329–335, 2002.

146. M. Gillet, K. Aguir, M. Bendahan, and P. Mennini, Grain size effect in sputtered tungsten trioxide thin films on the sensitivity to ozone, *Thin Solid Films*, **484**, 358–363, 2005.

147. A. Ponzoni, E. Comini, M. Ferroni, and G. Sberveglieri, Nanostructured WO$_3$ deposited by modified thermal evaporation for gas-sensing applications, *Thin Solid Films*, **490**, 81–85, 2005.

148. Y.-G. Choi, G. Sakai, K. Shimanoe, Y. Teraoka, N. Miura, and N. Yamazoe, Preparation of size and habit-controlled nano crystallites of tungsten oxide, *Sensors and Actuators*, **B93**, 486–494, 2003.

149. Y.-G. Choi, G. Sakai, K. Shimanoe, N. Miura, and N. Yamazoe, Preparation of aqueous sols of tungsten oxide dihydrate from sodium tungstate by an ion-exchange method, *Sensors and Actuators*, **B87**, 63–72, 2002.

150. S.-H. Wang, T.-C. Chou, and C.-C. Liu, Nano-crystalline tungsten oxide NO_2 sensor, *Sensors and Actuators*, **B94**, 343–351, 2003.

151. X. Li, G. Zhang, F. Cheng, B. Guo, and J. Chen, Synthesis, characterization, and gas-sensor application of WO_3 nanocuboids, *Journal of the Electrochemical Society*, **153**, H133–H137, 2006, and the references therein.

152. X.-L. Li, T.-J. Lou, X.-M. Sun, and Y.-D. Li, Highly sensitive WO_3 hollow-sphere gas sensors, *Inorganic Chemistry*, **43**, 5442–5449, 2004.

153. E. Rossinyol, A. Marsal, J. Arbiol, F. Peiró, A. Cornet, J.R. Morante, B. Tian, T. Bo, and D. Zhao, Crystalline mesoporous tungsten oxide for gas sensing applications, *IEEE Proceedings of the Spanish Conference on Electron Devices*, **5**, 517–520, 2005.

154. A. Ponzoni, E. Comini, G. Sberveglieri, J. Zhou, S.Z. Deng, N.S. Xu, Y. Ding, and Z.L. Wang, Ultrasensitive and highly selective gas sensors using three-dimensional tungsten oxide nanowire networks, *Applied Physics Letters*, **88**, 203101, 2006.

155. W. Chen, C. Xiao, Q. Yang, S. Yang, and A. Hirose, Controlled synthesis of tungsten and tungsten oxide nanorod films, *Materials Research Innovations Online*, **10** (2), 1433-075X, 169–178, 2006.

156. G. Gu, B. Zheng, W.Q. Han, S. Roth, and J. Liu, Tungsten oxide nanowires on tungsten substrates, *Nano Letters*, **2**, 849–851, 2002.

157. I. Jiménez, A.M. Vilà, A.C. Calveras, and J.R. Morante, Gas-sensing properties of catalytically modified WO_3 with copper and vanadium for NH_3 detection, *IEEE Sensors Journal*, **5**, 385–391, 2005.

158. V. Khatko, E. Llobet, X. Vilanova, J. Brezmes, J. Hubalek, K. Malysz, and X. Correig, Gas sensing properties of nanoparticle indium-doped WO_3 thick films, *Sensors and Actuators*, **B111–B112**, 45–51, 2005.

159. A. Hoel, L.F. Reyes, S. Saukko, P. Heszler, V. Lantto, and C.G. Granqvist, Gas sensing with films of nanocrystalline WO_3 and Pd made by advanced reactive gas deposition, *Sensors and Actuators*, **B105**, 283–289, 2005.

160. H.Y. Yu, B.H. Kang, U.H. Pi, C.W. Park, S.-Y. Choi, and G.T. Kim, V_2O_5 nanowire-based nanoelectronic devices for helium detection, *Applied Physics Letters*, **86**, 253102, 2005.

161. Ch. Pandis, N. Brilis, E. Bourithis, D. Tsamakis, H. Ali, S. Krishnamoorthy, A.A. Iliadis, and M. Kompitsas, Low-temperature hydrogen sensors based on Au nanoclusters and Schottky contacts on ZnO films deposited by pulsed laser deposition on Si and SiO_2 substrates, *IEEE Sensors Journal*, **7**, 448–454, 2007.

162. S. Kim, H. Moon, D. Gupta, S. Yoo, and Y.-K. Choi, Resistive switching characteristics of sol–gel zinc oxide films for flexible memory applications, *IEEE Transactions on Electron Devices*, **56**, 696–682, 2007.

163. J. Yang, G. Liu, J. Lu, Y. Qiu, and S. Yang, Electrochemical route to the synthesis of ultrathin ZnO nanorod/nanobelt arrays on zinc substrate, *Applied Physics Letters*, **90**, 103109, 2007.

164. S. Choopun, N. Hongsith, P. Mangkorntong, and N. Mangkorntong, Zinc oxide nanobelts by RF sputtering for ethanol sensor, *Physica E: Low-Dimensional Systems and Nanostructures*, **39**, 53–56, 2007.

165. Y.W. Heo, L.C. Tien, D.P. Norton, B.S. Kang, F. Ren, B.P. Gila, and S.J. Pearton, Electrical transport properties of single ZnO nanorods, *Applied Physics Letters*, **85**, 2002–2004, 2004.

166. J.F. Conley, Jr., L. Stecker, and Y. Ono, Directed integration of ZnO nanobridge devices on a Si substrate, *Applied Physics Letters*, **87**, 223114, 2005.

167. C.C. Li, Z.F. Du, L.M. Li, H.C. Yu, Q. Wan, and T.H. Wang, Surface-depletion controlled gas sensing of ZnO nanorods grown at room temperature, *Applied Physics Letters*, **91**, 032101, 2007.

168. A. Khan, W.M. Jadwisienczak, H.J. Lozykowski, and M.E. Kordesch, Catalyst-free synthesis and luminescence of aligned ZnO nanorods, *Physica E: Low-Dimensional Systems and Nanostructures*, **39**, 258–261, 2007.

169. P. Feng, Q. Wan, and T.H. Wang, Contact-controlled sensing properties of flowerlike ZnO nanostructures, *Applied Physics Letters*, **87**, 213111, 2005, and the references therein.

170. S.-W. Kim, H.-K. Park, M.-S. Yi, N.-M. Park, J.-H. Park, S.-H. Kim, S.-L. Maeng, C.-J. Choi, and S.-E. Moon, Epitaxial growth of ZnO nanowall networks on GaN/sapphire substrates, *Applied Physics Letters*, **90**, 033107, 2007.

171. T.-J. Hsueh, Y.-W. Chen, S.-J. Chang, S.-F. Wang, C.-L. Hsu, Y.-R. Lin, T.-S. Lin, and I.-C. Chen, ZnO nanowire-based CO sensors prepared at various temperatures, *Journal of the Electrochemical Society*, **154**, J393–J396, 2007.

172. K.A. Jeon, H.J. Son, C.E. Kim, J.H. Kim, and S.Y. Lee, Photoluminescence of ZnO nanowires grown on sapphire (11$\bar{2}$0) substrates, *Physica E: Low-Dimensional Systems and Nanostructures*, **37**, 222–225, 2007.

173. Z. Fan, D. Wang, P.-C. Chang, W.-Y. Tseng, and J.G. Lu, ZnO nanowire field-effect transistor and oxygen sensing property, *Applied Physics Letters*, **85**, 5923–5925, 2004.

174. Q.H. Li, Y.X. Liang, Q. Wan, and T.H. Wang, Oxygen sensing characteristics of individual ZnO nanowire transistors, *Applied Physics Letters*, **85**, 6389–6391, 2004.

175. R. Tena-Zaera, M.A. Ryan, A. Katty, G. Hodes, S. Bastide, and C. Lévy-Clément, Fabrication and characterization of ZnO nanowires/CdSe/CuSCN *eta*-solar cell, *Comptes Rendus Chimie*, **9**, 717–729, 2006.

176. T.-J. Hsueh, S.-J. Chang, C.-L. Hsu, Y.-R. Lin, and I.-C. Chen, Highly sensitive ZnO nanowire ethanol sensor with Pd adsorption, *Applied Physics Letters*, **91**, 053111, 2007, and the references therein.

177. S. Ren, Y.F. Bai, J. Chen, S.Z. Deng, N.S. Xu, Q.B. Wu, and S. Yang, Catalyst-free synthesis of ZnO nanowire arrays on zinc substrate by low temperature thermal oxidation, *Materials Letters*, **61**, 666–670, 2007.

178. W. Wang, G. Zhang, L. Yu, X. Bai, Z. Zhang, and X. Zhao, Field emission properties of zinc oxide nanowires fabricated by thermal evaporation, *Physica E: Low-Dimensional Systems and Nanostructures*, **36**, 86–91, 2007.

179. S.J. Young, L.W. Ji, S.J. Chang, T.H. Fang, and T.J. Hsueh, Nanoindentation of vertical ZnO nanowires, *Physica E: Low-Dimensional Systems and Nanostructures*, **39**, 240–243, 2007, and the references therein.

180. C.-Y. Lu, S.-P. Chang, S.-J. Chang, T.-J. Hsueh, C.-L. Hsu, Y.-Z. Chiou, and I.-C. Chen, ZnO nanowire-based oxygen gas sensor, *IEEE Sensors Journal*, **9**, 485–489, 2009.

181. H.T. Ng, J. Han, T. Yamada, P. Nguyen, Y.P. Chen, and M. Meyyappan, Single crystal nanowire vertical surround-gate field-effect transistor, *Nano Letters*, **4**, 1247–1252, 2004.

182. N. Takahashi, Simple and rapid synthesis of ZnO nano-fiber by means of a domestic microwave oven, *Materials Letters*, **62**, 1652–1654, 2008.

183. B.J. Chen, X.W. Sun, C.X. Xu, and B.K. Tay, Growth and characterization of zinc oxide nano/micro-fibers by thermal chemical reactions and vapor transport deposition in air, *Physica E: Low-Dimensional Systems and Nanostructures*, **21**, 103–107, 2004.

184. R. Maity, S. Das, M.K. Mitra, and K.K. Chattopadhyay, Synthesis and characterization of ZnO nano/microfibers thin films by catalyst free solution route, *Physica E: Low-Dimensional Systems and Nanostructures*, **25**, 605–612, 2005.

185. H.T. Wang, B.S. Kang, F. Ren, L.C. Tien, P.W. Sadik, D.P. Norton, S.J. Pearton, and J. Lin, Hydrogen-selective sensing at room temperature with ZnO nanorods, *Applied Physics Letters*, **86**, 243503, 2005.

186. Y.W. Heo, D.P. Norton, L.C. Tien, Y. Kwon, B.S. Kang, F. Ren, S.J. Pearton, and J.R. LaRoche, ZnO nanowire growth and devices, *Materials Science and Engineering*, **R47**, 1–47, 2004, and the references therein.

187. Y. Lv, L. Guo, H. Xu, and X. Chu, Gas-sensing properties of well-crystalline ZnO nanorods grown by a simple route, *Physica E: Low-Dimensional Systems and Nanostructures*, **36**, 102–105, 2007.

188. C. Wang, X. Chu, and M. Wu, Detection of H_2S down to ppb levels at room temperature using sensors based on ZnO nanorods, *Sensors and Actuators*, **B113**, 320–323, 2006.

189. B.I. Seo, U.A. Shaislamov, M.H. Ha, S.-W. Kim, H.-K. Kim, and B. Yang, ZnO nanotubes by template wetting process, *Physica E: Low-Dimensional Systems and Nanostructures*, **37**, 241–244, 2007.

190. H. Gong, J.Q. Hu, J.H. Wang, C.H. Ong, and F.R. Zhu, Nano-crystalline Cu-doped ZnO thin film gas sensor for CO, *Sensors and Actuators*, **B115**, 247–251, 2006.

191. N. Wang, C.G. Hu, C.H. Xia, B. Feng, Z.W. Zhang, Y. Xi, and Y.F. Xiong, Ultrasensitive gas sensitivity property of $BaMnO_3$ nanorods, *Applied Physics Letters*, **90**, 163111, 2007.

192. J. Yuh, L. Perez, W.M. Sigmund, and J.C. Nino, Electrospinning of complex oxide nanofibers, *Physica E: Low-Dimensional Systems and Nanostructures*, **37**, 254–259, 2007.

193. R. Ferro, J.A. Rodríguex, I. Jiménez, A. Cirera, J. Cerdà, and J.R. Morante, Gas-sensing properties of sprayed films of $(CdO)_x(ZnO)_{1-x}$ mixed oxide, *IEEE Sensors Journal*, **5**, 48–52, 2005.

194. N. Izu, N. Oh-hori, M. Itou, W. Shin, I. Matsubara, and N. Murayama, Resistive oxygen gas sensors based on $Ce_{1-x}Zr_xO_2$ nano powder prepared using new precipitation method, *Sensors and Actuators*, **B108**, 238–243, 2005.

195. T. Jantson, T. Avarmaa, H. Mändar, T. Uustare, and R. Jaaniso, Nanocrystalline Cr_2O_3–TiO_2 thin films by pulsed laser deposition, *Sensors and Actuators*, **B109**, 24–31, 2005.

196. M. Ivanovskaya, D. Kotsikau, G. Faglia, P. Nelli, and S. Irkaev, Gas-sensitive properties of thin film heterojunction structures based on Fe_2O_3–In_2O_3 nanocomposites, *Sensors and Actuators*, **B93**, 422–430, 2003.

197. M.I. Ivanovskaya, D.A. Kotsikau, A. Taurino, and P. Siciliano, Structural distinctions of Fe_2O_3–In_2O_3 composites obtained by various sol–gel procedures, and their gas-sensing features, *Sensors and Actuators*, **B124**, 133–142, 2007.

198. O.K. Tan, W. Cao, W. Zhu, J.W. Chai, and J.S. Pan, Ethanol sensors based on nano-sized α-Fe_2O_3 with SnO_2, ZrO_2, TiO_2 solid solutions, *Sensors and Actuators*, **B93**, 396–401, 2003.

199. M.N. Rumyantseva, V.V. Kovalenko, A.M. Gaskov, T. Pagnier, D. Machon, J. Arbiol, and J.R. Morante, Nanocomposites SnO_2/Fe_2O_3: Wet chemical synthesis and nanostructure characterization, *Sensors and Actuators*, **B109**, 64–74, 2005.

200. G. Zhang and J. Chen, Synthesis and application of $La_{0.59}Ca_{0.41}CoO_3$ nanotubes, *Journal of the Electrochemical Society*, **152**, A2069–A2073, 2005.

201. G.-Y. Zhang, B. Guo, and J. Chen, MCo_2O_4 (M = Ni, Cu, Zn) nanotubes: Template synthesis and application in gas sensors, *Sensors and Actuators*, **B114**, 402–409, 2006.

202. Y. Hu, O.K. Tan, J.S. Pan, H. Huang, and W. Cao, The effects of annealing temperature on the sensing properties of low temperature nano-sized $SrTiO_3$ oxygen gas sensor, *Sensors and Actuators*, **B108**, 244–249, 2005.

203. Y. Hu, O.K. Tan, W. Cao, and W. Zhu, Fabrication and characterization of nano-sized $SrTiO_3$-based oxygen sensor for near room-temperature operation, *IEEE Sensors Journal*, **5**, 825–832, 2005.

204. M. Ferroni, V. Guidi, G. Martinelli, P. Nelli, and G. Sberveglieri, Gas-sensing applications of W–Ti–O-based nanosized thin films prepared by r.f. reactive sputtering, *Sensors and Actuators*, **B44**, 499–502, 1997.

205. C.V.G. Reddy, W. Cao, O.K. Tan, W. Zhu, and S.A. Akbar, Selective detection of ethanol vapor using $xTiO_2$–(1 − x)WO_3 based sensor, *Sensors and Actuators*, **B94**, 99–102, 2003.

206. O.K. Tan, W. Cao, and W. Zhu, Alcohol sensor based on a non-equilibrium nano-structured $xZrO_2$–(1 − x)α-Fe_2O_3 solid solution system, *Sensors and Actuators*, **B63**, 129–134, 2000.

207. A. Cirera, A. Vilà, A. Diéguez, A. Cabot, A. Cornet, and J.R. Morante, Microwave processing for the low cost, mass production of undoped and in situ catalytic doped nanosized SnO_2 gas sensor powders, *Sensors and Actuators*, **B64**, 65–69, 2000.

208. D.B. Allred, M.T. Zin, H. Ma, M. Sarikaya, F. Baneyx, A.K.-Y. Jen, and D.T. Schwartz, Direct nanofabrication and transmission electron microscopy on a suite of easy-to-prepare ultrathin film substrates, *Thin Solid Films*, **515**, 5341–5347, 2007.

209. M. Kang, D. Kim, S.H. Yi, J.U. Han, J.E. Yie, and J.M. Kim, Preparation of stable mesoporous inorganic oxides via nano-replication technique, *Catalysis Today*, **93–95**, 695–699, 2004.

210. T. Seiyama and S. Kagawa, Study on a detector for gaseous components using semiconductive thin films, *Analytical Chemistry*, **38**, 1069–1073, 1966.

211. V.E. Henrich and P.A. Cox, *The Surface Science of Metal Oxides*, Cambridge University Press, Cambridge, U.K., 1994, and the references therein.

212. T. Okabayashi, T. Fujimoto, I. Yamamoto, K. Utsunomiya, T. Wada, Y. Yamashita, N. Yamashita, and M. Nakagawa, High sensitive hydrocarbon gas sensor utilizing cataluminescence of γ-Al_2O_3 activated with Dy^{3+}, *Sensors and Actuators*, **B64**, 54–58, 2000.

213. K.T. Kim, S.J. Sim, and S.M. Cho, Hydrogen gas sensor using Pd nanowires electro-deposited into anodized alumina template, *IEEE Sensors Journal*, **6**, 509–513, 2006.

214. Z.N. Adamian, H.V. Abovian, and V.M. Aroutiounian, Smoke sensor on the base of Bi_2O_3 sesquioxide, *Sensors and Actuators*, **B35–B36**, 241–243, 1996.

215. G.S. Devi, S.V. Manorama, and V.J. Rao, SnO_2/Bi_2O_3: A suitable system for selective carbon monoxide detection, *Journal of the Electrochemical Society*, **145**, 1039–1044, 1998.

216. M. Wark, H. Kessler, and G. Schulz-Ekloff, Growth and reactivity of zinc and cadmium oxide nano-particles in zeolites, *Microporous Materials*, **8**, 241–253, 1997.

217. G. Sberveglieri, S. Groppelli, and P. Nelli, Highly sensitive and selective NO_x and NO_2 sensor based on Cd-doped SnO_2 thin films, *Sensors and Actuators*, **B4**, 457–461, 1991.

218. N. Izu, W. Shin, I. Matsubara, and N. Murayama, The effects of the particle size and crystallite size on the response time for resistive oxygen gas sensor using cerium oxide thick film, *Sensors and Actuators*, **B94**, 222–227, 2003.

219. Z. Mei, W. Xidong, W. Fuming, and L. Wenchao, Oxygen sensitivity of nano-CeO_2 coating TiO_2 materials, *Sensors and Actuators*, **B92**, 167–170, 2003.

220. A. Chowdhuri, V. Gupta, K. Sreenivas, R. Kumar, S. Mozumdar, and P.K. Patanjali, Response speed of SnO_2-based H_2S gas sensors with CuO nanoparticles, *Applied Physics Letters*, **84**, 1180–1182, 2004.

221. M. Fleischer and H. Meixner, Fast gas sensors based on metal oxides which are stable at high temperatures, *Sensors and Actuators*, **B43**, 1–10, 1997.

222. J. Frank, M. Fleischer, and H. Meixner, Gas-sensitive electrical properties of pure and doped semiconducting Ga_2O_3 thick films, *Sensors and Actuators*, **B48**, 318–321, 1998.

223. S. Jin, T.-W. Tang, and M.V. Fischetti, Simulation of silicon nanowire transistors using Boltzmann transport equation under relaxation time approximation, *IEEE Transactions on Electron Devices*, **55**, 727–736, 2008.

224. Y. Zheng, C. Rivas, R. Lake, K. Alam, T.B. Boykin, and G. Klimeck, Electronic properties of silicon nanowires, *IEEE Transactions on Electron Devices*, **52**, 1097–1103, 2005.

225. S. Poli, M.G. Pala, and T. Poiroux, Full quantum treatment of remote coulomb scattering in silicon nanowire FETs, *IEEE Transactions on Electron Devices*, **56**, 1191–1198, 2009.

226. Ch.Y. Wang, V. Cimalla, Th. Kups, C.-C. Röhlig, Th. Stauden, O. Ambacher, M. Kunzer, T. Passow, W. Schirmacher, W. Pletschen, K. Köhler, and J. Wagner, Integration of In_2O_3 nanoparticle based ozone sensors with GaInN/GaN light emitting diodes, *Applied Physics Letters*, **91**, 103509, 2007.

227. G. Neri, A. Bonavita, G. Micali, G. Rizzo, N. Pinna, and M. Niederberger, In_2O_3 and $Pt\text{-}In_2O_3$ nanopowders for low temperature oxygen sensors, *Sensors and Actuators*, **B127**, 455–462, 2007, and the references therein.

228. Y. Nakatani, M. Matsuoka, and Y. Iida, $\gamma\text{-}Fe_2O_3$ ceramic gas sensor, *IEEE Transactions on Components, Hybrids, and Manufacturing Technology*, **CHMT-5**, 522–527, 1982.

229. O.K. Tan, W. Zhu, Q. Yan, and L.B. Kong, Size effect and gas sensing characteristics of nanocrystalline $xSnO_2\text{--}(1 - x)\alpha\text{-}Fe_2O_3$ ethanol sensors, *Sensors and Actuators*, **B65**, 361–365, 2000.

230. K. Galatsis, Y.X. Li, W. Wlodarski, E. Comini, G. Sberveglieri, C. Cantalini, S. Santucci, and M. Passacantando, Comparison of single and binary oxide MoO_3, TiO_2 and WO_3 sol–gel gas sensors, *Sensors and Actuators*, **B83**, 276–280, 2002, and the references therein.

231. L. Chambon, C. Maleysson, A. Pauly, J.P. Germain, V. Demarne, and A. Grisel, Investigation, for NH_3 gas sensing applications, of the Nb_2O_5 semiconducting oxide in the presence of interferent species such as oxygen and humidity, *Sensors and Actuators*, **B45**, 107–114, 1997.

232. H. Windischmann and P. Mark, A model for the operation of a thin-film SnO$_x$ conductance-modulation carbon monoxide sensor, *Journal of the Electrochemical Society*, **126**, 627–633, 1979.

233. L.N. Yannopoulos, A *p*-type semiconductor thick film gas sensor, *Sensors and Actuators*, **12**, 263–273, 1987.

234. I.T. Weber, A. Valentini, L.F.D. Probst, E. Longo, and E.R. Leite, Influence of noble metals on the structural and catalytic properties of Ce-doped SnO$_2$ systems, *Sensors and Actuators*, **B97**, 31–38, 2004.

235. M. Batzill and U. Diebold, The surface and materials science of tin oxide, *Progress in Surface Science*, **79**, 47–154, 2005.

236. Y.J. Chen, L. Nie, X.Y. Xue, Y.G. Wang, and T.H. Wang, Linear ethanol sensing of SnO$_2$ nanorods with extremely high sensitivity, *Applied Physics Letters*, **88**, 083105, 2006.

237. Y.J. Chen, X.Y. Xue, Y.G. Wang, and T.H. Wang, Synthesis and ethanol sensing characteristics of single crystalline SnO$_2$ nanorods, *Applied Physics Letters*, **87**, 233503, 2005.

238. S.-D. Han, H. Yang, L. Wang, and J.-W. Kim, Preparation and properties of vanadium-doped SnO$_2$ nanocrystallites, *Sensors and Actuators*, **B66**, 112–115, 2000.

239. D. Wang, J. Jin, D. Xia, Q. Ye, and J. Long, The effect of oxygen vacancies concentration to the gas-sensing properties of tin dioxide-doped Sm, *Sensors and Actuators*, **B66**, 260–262, 2000.

240. J.R. Yu, G.Z. Huang, and Y.J. Yang, The effect of gas concentration on optical gas sensing properties of Ag- and Pt-doped SnO$_2$ ultrafine particle films, *Sensors and Actuators*, **B66**, 286–288, 2000.

241. A. Karthigeyan, R.P. Gupta, K. Scharnagl, M. Burgmair, M. Zimmer, S.K. Sharma, and I. Eisele, Low temperature NO$_2$ sensitivity of nano-particulate SnO$_2$ film for work function sensors, *Sensors and Actuators*, **B78**, 69–72, 2001.

242. C. Yu, Q. Hao, S. Saha, L. Shi, X. Kong, and Z.L. Wang, Integration of metal oxide nanobelts with microsystems for nerve agent detection, *Applied Physics Letters*, **86**, 063101, 2005.

243. Z.W. Pan, Z.R. Dai, and Z.L. Wang, Nanobelts of semiconducting oxides, *Science*, **291**, 1947–1949, 2001.

244. A. Helwig, G. Müller, M. Eickhoff, and G. Sberveglieri, Dissociative gas sensing at metal oxide surfaces, *IEEE Sensors Journal*, **7**, 1675–1679, 2007, and the references therein.

245. B.-U. Moon, J.-M. Lee, C.-H. Shim, M.-B. Lee, J.-H. Lee, D.-D. Lee, and J.-H. Lee, Silicon bridge type micro-gas sensor array, *Sensors and Actuators*, **B108**, 271–277, 2005.

246. D. Barrettino, M. Graf, S. Taschini, S. Hafizovic, C. Hagleitner, and A. Hierlemann, CMOS monolithic metal-oxide gas sensor microsystems, *IEEE Sensors Journal*, **6**, 276–286, 2006, and the references therein.

247. G. Sberveglieri, S. Groppelli, P. Nelli, and A. Camanzi, A new technique for the preparation of highly sensitive hydrogen sensors based on SnO$_2$(Bi$_2$O$_3$) thin films, *Sensors and Actuators*, **B5**, 253–255, 1991.

248. J.C. Xu, G.W. Hunter, D. Lukco, C.-C. Liu, and B.J. Ward, Novel carbon dioxide microsensor based on tin oxide nanomaterial doped with copper oxide, *IEEE Sensors Journal*, **9**, 235–236, 2009.

249. K.-i. Ishibashi, R.-t. Yamaguchi, Y. Kimura, and M. Niwano, Fabrication of titanium oxide nanotubes by rapid and homogeneous anodization in perchloric acid/ethanol mixture, *Journal of the Electrochemical Society*, **155**, K10–K14, 2008.

250. Y. Xu, X. Zhou, and O.T. Sorensen, Oxygen sensors based on semiconducting metal oxides: An overview, *Sensors and Actuators*, **B65**, 2–4, 2000, and the references therein.

251. M. Ferroni, M.C. Carotta, V. Guidi, G. Martinelli, F. Ronconi, O. Richard, D.V. Dyck, and J.V. Landuyt, Structural characterization of Nb-TiO$_2$ nanosized thick-films for gas sensing application, *Sensors and Actuators*, **B68**, 140–145, 2000.

252. Z. Wang, E. Han, and W. Ke, Effect of nanoparticles on the improvement in fire-resistant and anti-ageing properties of flame-retardant coating, *Surface and Coatings Technology*, **200**, 5706–5716, 2006.

253. L. Francioso, M. Prato, and P. Siciliano, Low-cost electronics and thin film technology for sol–gel titania lambda probes, *Sensors and Actuators*, **B128**, 359–365, 2008.

254. L.E. Depero, M. Ferroni, V. Guidi, G. Marca, G. Martinelli, P. Nelli, L. Sangaletti, and G. Sberveglieri, Preparation and micro-structural characterization of nanosized thin film of TiO$_2$-WO$_3$ as a novel material with high sensitivity towards NO$_2$, *Sensors and Actuators*, **B35–B36**, 381–383, 1996.

255. O. Berger, W.-J. Fischer, and V. Melev, Tungsten-oxide thin films as novel materials with high sensitivity and selectivity to NO$_2$, O$_3$ and H$_2$S. Part-I: Preparation and microstructural characterization of the tungsten-oxide thin films, *Journal of Materials Science: Materials in Electronics*, **15**, 463–482, 2004.

256. K. Lee, W.S. Seo, and J.T. Park, Synthesis and optical properties of colloidal tungsten oxide nanorods, *Journal of American Chemical Society*, **125**, 3408–3409, 2003.

257. S. Surnev, M.G. Ramsey, and F.P. Netzer, Vanadium oxide surface studies, *Progress in Surface Science*, **73**, 117–165, 2003.

258. X. Fang, Y. Bando, U.K. Gautam, T. Zhai, H. Zeng, X. Xu, M. Liao, and D. Golberg, ZnO and ZnS nanostructures: Ultraviolet-light emitters, lasers, and sensors, *Critical Reviews in Solid State and Materials Sciences*, **34**, 190–223, 2009.

259. V. Avrutin, G. Cantwell, J. Zhang, J.J. Song, D.J. Silversmith, and H. Morkoç, Bulk ZnO: Current status, challenges, and prospects, *Proceedings of the IEEE*, **98**, 1339–1350, 2010.

260. J. Xu, Q. Pan, Y. Shun, and Z. Tian, Grain size control and gas sensing properties of ZnO gas sensor, *Sensors and Actuators*, **B66**, 277–279, 2000.

261. M. Driess, K. Merz, R. Schoenen, S. Rabe, F.E. Kruis, A. Roy, and A. Birkner, From molecules to metastable solids: Solid-state and chemical vapour syntheses (CVS) of nanocrystalline ZnO and Zn, *Comptes Rendus Chimie*, **6**, 273–281, 2003, and the references therein.

262. G. Socol, E. Axente, C. Ristoscu, F. Sima, A. Popescu, N. Stefan, I.N. Mihailescu, L. Escoubas, J. Ferreira, S. Bakalova, and A. Szekeres, Enhanced gas sensing of Au nanocluster-doped or -coated zinc oxide thin films, *Journal of Applied Physics*, **102**, 083103, 2007.

263. X.Y. Xue, Y.J. Chen, Y.G. Wang, and T.H. Wang, Synthesis and ethanol sensing properties of ZnSnO$_3$ nanowires, *Applied Physics Letters*, **86**, 233101, 2005.

264. X.Y. Xue, P. Feng, Y.G. Wang, and T.H. Wang, Extremely high oxygen sensing of individual ZnSnO$_3$ nanowires arising from grain boundary barrier modulation, *Applied Physics Letters*, **91**, 022111, 2007.

265. J. Wang, B. Xu, G. Liu, J. Zhang, and T. Zhang, Improvement of nanocrystalline BaTiO$_3$ humidity sensing properties, *Sensors and Actuators*, **B66**, 159–160, 2000.

266. I. Kosacki and H.U. Anderson, Nanostructured oxide thin films for gas sensors, *Sensors and Actuators*, **B48**, 263–269, 1998.

267. Y. Hu, O.K. Tan, W. Cao, and W. Zhu, A low temperature nano-structured SrTiO₃ thick film oxygen gas sensor, *Ceramics International*, **30**, 1819–1822, 2004, and the references therein.

268. D.H. Yun, C.H. Kwon, H.-K. Hong, S.-R. Kim, K. Lee, H.G. Song, and J.E. Kim, Highly sensitive ammonia gas sensor, in *Proceedings of the 1997 International Conference on Solid-State Sensors and Actuators*, Chicago, IL, June 16–19, 959–962, 1997.

269. A. Trinchi, K. Galatsis, W. Wlodarski, and Y.X. Li, A Pt/Ga_2O_3-ZnO/SiC Schottky diode-based hydrocarbon gas sensor, *IEEE Sensors Journal*, **3**, 548–553, 2003.

270. A. Chiorino, G. Ghiotti, F. Prinetto, M.C. Carotta, M. Gallana, and G. Martinelli, Characterization of materials for gas sensors. Surface chemistry of SnO_2 and MoO_x–SnO_2 nano-sized powders and electrical responses of the related thick films, *Sensors and Actuators*, **B59**, 203–209, 1999.

271. J. Esch, The future of integrated circuits: A survey of nanoelectronics, *Proceedings of the IEEE*, **98**, 8–10, 2010.

272. M. Haselman and S. Hauck, The future of integrated circuits: A survey of nano-electronics, *Proceedings of the IEEE*, **98**, 11–38, 2010.

273. R. Kim and M.S. Lundstrom, Physics of carrier backscattering in one- and two-dimensional nanotransistors, *IEEE Transactions on Electron Devices*, **56**, 132–139, 2009, and the references therein.

274. M.S. Arnold, P. Avouris, Z.W. Pan, and Z.L. Wang, Field-effect transistors based on single semiconducting oxide nanobelts, *Journal of Physical Chemistry*, **B107**, 659–663, 2003.

275. H. Morkoç, J.-I. Chyi, A. Krost, Y. Nanishi, and D.J. Silversmith, Challenges and opportunities in GaN and ZnO devices and materials, *Proceedings of the IEEE*, **98**, 1113–1117, 2010.

276. Ü. Özgür, D. Hofstetter, and H. Morkoç, ZnO devices and applications: A review of current status and future prospects, *Proceedings of the IEEE*, **98**, 1255–1268, 2010, and the references therein.

277. Z. Fan and J.G. Lu, Gate-refreshable nanowire chemical sensors, *Applied Physics Letters*, **86**, 123510, 2005.

278. C. Chen, K. Abe, H. Kumomi, and J. Kanicki, Density of states of a-InGaZnO from temperature-dependent field-effect studies, *IEEE Transactions on Electron Devices*, **56**, 1177–1183, 2009.

279. R. Shimada and H. Morkoç, Wide bandgap semiconductor-based surface-emitting lasers: Recent progress in GaN-based vertical cavity surface-emitting lasers and GaN-/ZnO-based polariton lasers, *Proceedings of the IEEE*, **98**, 1113–1117, 2010.

280. R.B.M. Cross, M.M. De Souza, S.C. Deane, and N.D. Young, A comparison of the performance and stability of ZnO-TFTs with silicon dioxide and nitride as gate insulators, *IEEE Transactions on Electron Devices*, **55**, 1109–1115, 2008.

281. L. Shang, M. Liu, D. Tu, G. Liu, X. Liu, and Z. Ji, Low-voltage organic field-effect transistor with PMMA/ZrO_2 bilayer dielectric, *IEEE Transactions on Electron Devices*, **56**, 370–376, 2009.

282. M.G. McDowell and I.G. Hill, Influence of channel stoichiometry on zinc indium oxide thin-film transistor performance, *IEEE Transactions on Electron Devices*, **56**, 343–347, 2009.

283. L. Guo, P.R. Krauss, and S.Y. Chou, Nanoscale silicon field effect transistors fabricated using imprint lithography, *Applied Physics Letters*, **71**, 1881–1883, 1997.

284. D.K. Ferry, M.J. Gilbert, and R. Akis, Some considerations on nanowires in nanoelectronics, *IEEE Transactions on Electron Devices*, **55**, 2820–2826, 2008.

285. J. Appenzeller, J. Knoch, M.T. Björk, H. Riel, H. Schmid, and W. Riess, Toward nanowire electronics, *IEEE Transactions on Electron Devices*, **55**, 2827–2845, 2008.

286. E. Gnani, A. Gnudi, S. Reggiani, and G. Baccarani, Quasi-ballistic transport in nanowire field-effect transistors, *IEEE Transactions on Electron Devices*, **55**, 2918–2930, 2008.

287. W. Lu, P. Xie, and C.M. Lieber, Nanowire transistor performance limits and applications, *IEEE Transactions on Electron Devices*, **55**, 2859–2876, 2008.

288. J. Kong, N.R. Franklin, C. Zhou, M.G. Chapline, S. Peng, K. Cho, and H. Dai, Nanotube molecular wires as chemical sensors, *Science*, **287**, 622–625, 2000.

289. G. Fiori, G. Iannaccone, and G. Klimeck, A three-dimensional simulation study of the performance of carbon nanotube field-effect transistors with doped reservoirs and realistic geometry, *IEEE Transactions on Electron Devices*, **53**, 1782–1788, 2006.

290. A. Raychowdhury, A. Keshavarzi, J. Kurtin, V. De, and K. Roy, Carbon nanotube field-effect transistors for high-performance digital circuits—DC analysis and modeling toward optimum transistor structure, *IEEE Transactions on Electron Devices*, **53**, 2711–2717, 2006.

291. K. Rogdakis, S.-Y. Lee, M. Bescond, S.-K. Lee, E. Bano, and K. Zekentes, 3C-Silicon carbide nanowire FET: An experimental and theoretical approach, *IEEE Transactions on Electron Devices*, **55**, 1970–1976, 2008.

292. G. Liang, N. Neophytou, D.E. Nikonov, and M.S. Lundstrom, Performance projections for ballistic graphene nanoribbon field-effect transistors, *IEEE Transactions on Electron Devices*, **54**, 677–682, 2007.

293. J. Kedzierski, P.-L. Hsu, P. Healey, P.W. Wyatt, C.L. Keast, M. Sprinkle, C. Berger, and W.A. de Heer, Epitaxial graphene transistors on SiC substrates, *IEEE Transactions on Electron Devices*, **55**, 2078–2085, 2009.

294. T. Zeng and W. Liu, Phonon heat conduction in micro- and nano-core-shell structures with cylindrical and spherical geometries, *Journal of Applied Physics*, **93**, 4163–4168, 2003.

295. T. Kubo and H. Nozoye, Physical properties of spinel nano-structure epitaxially grown on MgO(100), *Applied Surface Science*, **188**, 545–549, 2002.

Index